BEEKEEPING

A Beginner's Guide to Building a Hive for Your Bee Colony

By

Jason Howard

© Copyright 2019 - All rights reserved.

The contents of this book may not be reproduced, duplicated or transmitted without direct written permission from the author.

Under no circumstances will any legal responsibility or blame be held against the publisher for any reparation, damages, or monetary loss due to the information herein, either directly or indirectly.

Legal Notice:

This book is copyright protected. This is only for personal use. You cannot amend, distribute, sell, use, quote or paraphrase any part or the content within this book without the consent of the author.

Disclaimer Notice:

Please note the information contained within this document is for educational and entertainment purposes only. Every attempt has been made to provide accurate, up to date and reliable complete information. No warranties of any kind are expressed or implied. Readers acknowledge that the author is not engaging in the rendering of legal, financial, medical or professional advice. The content of this book has been derived from various sources. Please consult a licensed professional before attempting any techniques outlined in this book.

BEEKEEPING

By reading this document, the reader agrees that under no circumstances are is the author responsible for any losses, direct or indirect, which are incurred as a result of the use of information contained within this document, including, but not limited to, —errors, omissions, or inaccuracies.

BEEKEEPING

Table of Contents

Introduction V

Chapter 1 - What Is Beekeeping? 1

Chapter 2 - A Look At Benefits And Challenges 17

Chapter 3 - Types Of Honeybees 29

Chapter 4 - Types Of Hives 40

Chapter 5 - The Role Of Honeybees 58

Chapter 6 - Hive Maintenance And Treatment 75

Chapter 7 - Harvesting The Honey 103

Chapter 8 - Common Mistakes 117

Conclusion 125

INTRODUCTION

This book offers you an introduction to the magical art of beekeeping. It is far from a totally comprehensive book. To incorporate the thousands of years of acquired beekeeping knowledge would require several volumes running to thousands of pages. What it does offer is an overview that you will be able to use as a roadmap to getting started as a beekeeper.

There is an incredibly long history behind what we today know as beekeeping. Our predecessors have been doing this for a long time, and each successive generation of beekeepers has built onto the knowledge that went before them. Much of that knowledge seems difficult to master, but as you progress through this book, you will discover that none of what you are required to grasp is really that difficult. What is more, you will find that this is a subject that is almost addictive in its ability to fascinate.

Each tiny bee will only produce one-tenth of a teaspoon of honey over the course of its life, and yet a well-husbanded hive will comfortably give you forty pounds of honey in a season. For every pound of honey the hive produces, 20,000,000 flowers will have been visited by bees. During those foraging voyages, pollen will have been carried from one flower to another in a life-sustaining process that provides man with one in three of every forkful of food he puts into his mouth.

Beekeeping is more than just a matter of harnessing nature and profiting from the ability to do so. It is also an act of natural stewardship. As a good beekeeper, you will develop a relationship with your bees that have positive implications that go way beyond just your desire for a few jars of honey. There has never been a time when bees have needed you more than they do now. There has never been a time when man has needed bees more than he does now.

BEEKEEPING

CHAPTER 1

WHAT IS BEEKEEPING?

Beekeeping is the art, or science, depending on how you view things, of rearing and keeping bees in manmade hives. It is a subject that can be so complicated that people devote their entire lives to its study and yet so simple that man has been able to practice it for thousands of years. Whether you want to start with just one hive in your back garden or you hope to have dozens of hives and turn beekeeping into a business, this book will walk you through the hows and whys of what you need to know. Good beekeeping involves husbandry and careful stewardship. There is a delicate balance to be maintained so that you can harvest some of the honey while at the same time leaving enough to maintain the hive in a state of good health, especially over the cold and lean winter months.

There are many factors that motivate people to start keeping bees. Obviously, having access to one's own honey is one of the more commonly recognized motivators, but there are plenty of others. Bees are crucial to the overall health of the planet because they are so essential to the pollination process. People will argue that there are plenty of other creatures that also perform a role in the carrying of pollen from one plant or flower to another, and therefore the bee is not as crucial as it is made out to be. People who use that argument are a little ill-informed. Of course, there are other pollinating animals and insects, and some plants are able to self-pollinate. As far as you and I are concerned, however, one in three commercially produced crops requires bees at some stage in its pollination process. That means the humble honeybee is responsible for one in three of every mouthful you put in your mouth. Though there are over twenty thousand different types of bees, only one of those is suitable for honey production, and that is the honeybee.

Many years ago, honey was the main reason why big honey farmers kept bees. In some countries, most notably the United States, hiring hives to farmers to improve pollination is now the primary reason why they stay in business, with the honey being just a secondary by-product. The almond industry of California relies almost entirely on bees that are trucked in during the flowering season. As far as pollination is concerned, even though you may lack a giant almond orchard, a hive of bees can do wonders in improving the pollination of flowering plants in the domestic garden.

Another specialty line in the area of beekeeping is the rearing of queens. Queens are the heart of the bee colony, and maintaining a healthy queen is critical to having a healthy hive. As you will learn, there are ways of the breeding queen, which can then be sold to other beekeepers. The character of the queen dictates the character of all the other bees in her colony, and if the queen is bad-tempered, the hove will be more aggressive then if the queen is docile. It is common for owners of a hive of vicious stinging bees to swap out the queen for a more manageable one.

Other products that are produced by bees include wax, propolis, bee pollen, and royal jelly. All of these products provide an incentive to keep bees. Although honey is, and probably always will be, the product that we first think of when we think of keeping bees, the lesser know bi-products become available to the beekeepers should he choose to harvest and process them.

In recent years there has been a large increase in the number of people who have turned their hands to beekeeping. Bee numbers have seen a dramatic decline that has drawn a lot of attention from the press, and many people who are concerned about the well being of the environment have taken to keeping a hive or two simply to try to address some of that decline. We will look into some of the possible reasons why we are seeing colony collapse later in the book, but you should know that even if you only introduce one small hive, you are doing something positive for the natural environment. An ecologically managed hive will boost the environment in which it is situated. The bees will have a positive effect on the local flora, even if it is just a domestic garden, and this, in turn, provides food and habitat for other insects and animals. As a benefit, you suddenly have access to a number of products, nearly all of which are proven to provide health benefits. When you look at things in that context, the idea of supporting a beehive seems almost irresistible.

At this point, it may be a good idea to take a brief look at registration. There is no one size fits all answer here as different states have different rules, and you may also be governed by city or suburban by laws. There is also a big difference in the requirements of the domestic beekeeper that wants just one or two hives and the professional who may well need his honey house certified.

For the smaller non-professional, a good place to start is at your local council, where you can ask if there are any restrictions. Generally, there are no or very few, but check, nonetheless. If there is no law pertaining to beekeeping, there is probably one that covers nuisance, which will quickly be employed should problems develop. The best thing you can do is to make sure that no problem develops in the first place. Once you have ascertained that the local laws are not going to hold you back, the next port of call should be your neighbors. Many people have a preconceived idea that bees are dangerous. If you are well prepared and can demonstrate that you know what you are doing and have taken all the necessary precautions, then you can avoid many of the problems they will try to present.

Keeping your bees behind a hedge or fence is one way to lower risks of unintended stings. This protects the hive from wind, keeps it out of sight, and means the bees must take off at above head height, thus lowering the small possibility of them flying into someone unintentionally. Ensuring that they have easy access to water near the hive will stop them from having to pop over to the neighbor's pool if they need a drink. One of the times when your bees are likely to get most agitated is when you harvest the honey. There are ways to minimize this, and we will get to those later in the book. Even if your neighbors or some unhelpful bi law prevent you from having a hive in your back garden, you are by no means defeated. There are plenty of people who would be delighted to have a managed hive or tow on their farm or in their garden. Generally, an arrangement is reached where you place your hive on their property, and they are given a share of the honey produced. Of course, the side benefit of all that pollination goes to them.

If your ambitions extend to more than just beekeeping for a hobby and your own supply of honey, then things become a little more complicated. Again these rules vary from state to state and country to country if you are in Europe. Your nearest agricultural department should quickly be able to direct you to whoever it is that controls and inspects apiculturists. We all tend to avoid government interference, but in this case, there are distinct advantages that come with the regulation. You may need to get your honey house approved, and this could even entail an inspection annually. In some areas, they will also want to inspect your hives. The reason for this is to control and eradicate the spread of disease. Later in this book, we will see that disease is a major problem, especially to the professional beekeeper, and these inspections should be seen as something to help you rather than rules simply thrown up to make your life more difficult. Another distinct advantage here is that the state apiary inspection officers are experts in their own right. They have access to all the latest research and can supply scientific papers, news, and sometimes even further training. If there is a spread of disease somewhere in your state, these are likely to be some of the first people to know about it, and that information will probably affect the way you treat your hives.

BEEKEEPING

Instead of seeing him as your enemy, look on your apiary inspection officer as a rich source of information. One other factor has recently come into play. In some parts of the US, where beekeeping is big business, theft of whole colonies has become quite widespread. Here too, the inspection officer is likely to be aware of these incidents before you are, and he can at least give you a heads up as to whether or not you need to augment your security.

You may think that beekeeping is too complicated for the amateur or that you require a great deal of space or expensive equipment. Neither of these is issues that need to be a stumbling block. Inner-city beekeeping has really taken off in recent years. Parks and private gardens are often rich in flowering plants and a great source of nectar for your urban bee. If you are wondering why you have never seen any of these hives where you are living, it could be because the beekeeper has stationed them on top of buildings. Many tower blocks have large expanses of roof space that are not utilized, and these roofs are perfect places to place hives where they will not be disturbed but from which the bees will have easy access to the cities many flowering food sources. There is a growing awareness among city councils of the threat posed by the indiscriminate use of pesticides. Many of them have gone pesticide-free, and with that threat, removed bees often thrive better in the cities than they do in farmland.

We have all grown up reading children's stories with cartoons depicting swarms of angry bees chasing some unfortunate person down the road. In fact, if you speak to people with no working knowledge of bees, they will almost definitely tell you that bees are dangerous. This is one myth that needs to be cleared up quickly. Even if you have only a tiny garden, you will be able to keep at least one hive of bees. Bees only attack if they believe that the colony is threatened, and as long as you don't go too close, they will go about their busy lives and pay you little attention. In fact, as they become used to you and understand that you do not pose any sort of threat, they will ignore you even if you come right up to the hive. This does not mean that you should not take sensible precautions. If you have small children or pets, then the precautions you are going to have to take will need to be more carefully thought out. A football kicked into a hive, or an overly curious dog could cause the bee to panic, and that will lead to problems as they attempt to protect the colony.

If you are serious about keeping bees, there is a wealth of information available to you, but it is difficult to beat the benefits of some hands-on experience. The first time you open a hive and find yourself surrounded by bees, and the very distinct smell that emanates from their colony, is one that you will not soon forget. The beekeeping world is one of the most generous with information that I have ever encountered. Beekeeping clubs are the backbone of the amateur beekeeping world, and there will almost definitely be one within reach of you. If not, contact your nearest local beekeeper and tell him you are thinking about trying your hand at bee husbandry. It is highly likely that he will show you the ropes.

I have often found myself wondering why beekeepers are so free with their information, and the conclusion I have reached is that instead of seeing the newcomer as a threat, he sees him as an accomplice in an effort to make the world a better place. Beekeeping clubs offer an invaluable source of information, and they will also have a working knowledge of the local environment in which you are planning to set up. Very often, they have a group discount for the buying of equipment, and, as much of what you will need for harvesting the honey is only used infrequently, the club often shares items like honey extractors. Many clubs go on to serve as a place for social interaction between likeminded people. Unfortunately, these days, many beekeepers are getting on in age. This can play to your advantage as they may well be looking for a keen new enthusiast to whom they can donate equipment, providing they think you are serious. My first hives were a gift from an old hand that was getting ready to retire. Those old-timers are an absolute treasure trove of information.

Man has had a working relationship with bees that goes back millennia. Thousands of years before sugar became the primary sweetener; we were smacking our lips with delight over the sweet sticky honey that was produced by these tiny little flying insects. In fact, sugar is a much more recent addition to our dietary pallet. Sugar was originally only found in Polynesia. We know that it was consumed there in 510 BC and that from there, it would eventually be transported to India. It was not until the Crusades in the 11th century AD that the first westerners encountered it, and it would be centuries later that it would finally reach Europe in 1099. For hundreds of years after that, it was a delicacy that was confined to the very rich. It was heavily taxed in the United Kingdom until the tax was abolished under Prime Minister Gladstone in 1871. At that stage, it first started to become available to the less well off, but still at a price that meant it was a luxury rather than the everyday item we eat so much of today. The even more modern sweetener, corn syrup, has really only seen widespread use in the last couple of decades as industrial food companies strive for cheaper and cheaper ways to satiate our elevated craving for sweetness. Before that, honey or syrup derived from things such as the maple tree were our only options.

BEEKEEPING

Even before we began to master the complicated but fascinating art of beekeeping, honey had already become recognized as a precious commodity. Of course, then we did not know how to harness that honey making potential of bees, and we instead had to pilfer supplies from hives we found in the wild. Painted onto walls of a cave in Spain, called la Cuerva de la Arana, is a picture depicting a honey hunting expedition in what is believed to be one of the earliest pictorial records of such an event. It dates back fifteen thousand years.

It is difficult to ascertain when man first began to understand that if he could sustain the stings of angry bees, then he stood to reap the warm, sticky reward of a sweetness he had never before encountered. Early hominids probably quickly learned that if they followed honey birds, they could lead you to a hive of bees. There are several different types of honey bird, and the fact that they will guide people to honey in the hope that they will gain a share of the spoils has been well documented. Experts believe that as far back as the Stone Age, early man and birds co-evolved, and this relationship began to develop. That period dates back to 1.9 million years.

In Myanmar, a bee was found petrified in amber, and that insect is almost identical to the modern honeybee we are now so familiar with. Scientists estimate the age of that petrified insect at 100 million years. The last dinosaurs became extinct approximately 66 million years ago, which means that the bee must have been playing its crucial role as a pollinator even then and that, in doing so, it would have heavily influenced the direction that the evolutionary process took. Many of the large dinosaurs lived on a vegetarian diet, and that would have been made possible because of the bees. It also shows that the bee somehow survived the calamity that wiped out the much larger dinosaurs.

The earliest written record of bees to date is made on clay tablets written in cuneiform dating back to 3000 BC. Cuneiform is one of the earliest known means of writing. Biblically honey is mentioned in the Old Testament frequently but most notably, when it is used to describe the many benefits offered by a move to the Promised Land. It is often associated with the word of God. In the New Testament, it was one of the main things that sustained the profit of John while he lived in the desert as a hermit.

Looting beehives in the wild and actually cultivating them on a sustainable basis are two very different subjects, and it is difficult to pinpoint an exact date when this conversion took place. The ability to access honey without actually destroying the hive was a massive leap forward in our relationship with bees. The ancient Egyptians definitely made beehives, and honey was found in the tombs of many of the pharaohs, including that of Tutankhamen. One of the interesting things about honey is its anti-bacterial properties. Some of the honey found in ancient Egyptian tombs was still edible. The first examples of keeping bees were in earthenware vessels in north Africa at around 9000BC, and by 3500 BC, the Egyptian understanding of the subject had developed to a level where they were able to transport hives to different regions and use smoke to aid in honey extraction. It would not be until the 1800s that Europe would develop their skills to that extent. The Egyptians used honey in their mummification processes and had discovered the medicinal benefits that it offered. There are papyrus texts describing the health and medicinal benefits of honey and propolis, and these two products were used in over nine hundred different remedies. In the Egyptian hieroglyphic alphabet, the bee is the symbol of royalty.

Like the Egyptians, the Chinese have a long-standing relationship with the honeybee. The first ruler of the Chinese Zhou dynasty led his troops into battle under a flag bearing a bee as its emblem. As in Egypt, here are many ancient Chinese medical texts where the use of honey is included.

Aristotle, the keen Greek observer of nature, described the bees' ability to collect nectar from flowers and return it to their hives though he misinterpreted what happened after that, and he described honey as a deposit of the atmosphere. Lycurgus, the ancient founder of Sparta, modeled his government on behavior he had learned from watching the workings of bees in their hive. Bee husbandry was highly valued by the Greeks, who were quick to include honey in many of their dishes, and they appointed overseers of their bee colonies who were men of high standing.

The famous Russian author Leo Tolstoy was a beekeeper. Napoleon Bonaparte's robe was decorated with bees for his coronation as emperor. Louis XII had hoped to include a beehive in his coat of arms, but this was rejected by the National Convention, stating that 'bees have a queen.'

In Israel, there have been remains discovered of what is thought to be one of the first large scale apiary operations which date back to the ninth or tenth century BC. The two hundred hives would have required a fairly sophisticated level of beekeeping skills to maintain.

In India, honey has long been used in a drink, which was known as madhuparka, which young men as they prepared to ask for a young woman's hand in marriage and to special guests. The Sanskrit word madhu means honey and lends itself to the Slavic word medhu, which in turn is the root of the word mead in English.

BEEKEEPING

We may have failed to pin down an exact moment in time when beekeeping first started, but it is obvious that the relationship between man and bee goes back a very long way. There is no other insect on earth which man has become nearly as adept at profiting from. So powerful is our relationship to the bee that they have entered into the mythology of many different cultures.

The San Bushmen who inhabits the Kalahari Desert tells of a bee that carried a mantis across a river. The effort cost the bee his life, but before dying, it deposited a bee in the mantis, which, when hatched, gave birth to the first human. The tears from the sun god Ra of Egyptian mythology are said to have turned into bees. The Greek god Aristaeus was considered the god of beekeeping. While pursuing his wife, Eurydice, he caused her to stand on a snake, and the ensuing bite killed her. This so enraged her sisters that they killed all of his bees in revenge. Aristaeus sort advice from Proteus, who told him to sacrifice four bulls and four cows to honor his deceased wife. From the rotting corpses of these beasts emerged hives of bees. This is vaguely similar to mention of an incident in the Bible where Samson killed a lion, which was later inhabited by a swarm of bees. He shared the honey with his parents and later used the incident as a riddle to trick the Philistines. His wife, a Philistine herself, persuaded him to tell her the answer to the riddle, which she then shared with her brothers, and thus his plan failed. It is a famous biblical story, and yet one where the presence of the bees gets lost amongst all the other events that take place.

If the bee holds an important place in both history and mythology, it is nothing compared to the importance of the role they play today. Bees are major pollinators of many of the commercial crops that we eat on a daily basis, and even though honey is a wonderful product to eat, that pollinating role is far more important to man. We can all play a role in protecting these creatures, and with the current problems they face, perhaps it is time we did so.

CHAPTER 2

A LOOK AT BENEFITS AND CHALLENGES

Bees are under huge pressure at the moment. For reasons we do not yet fully understand, there has been a dramatic and rapid loss of bee colonies in many countries throughout the world. Colony collapse disorder, as it is frequently referred to, presents us with a rather frightening possibility. Not only do we stand to lose one of our primary pollinators, but we also risk doing massive damage to human society in the process. In 2019 both the Royal Geographic Society and Earthwatch both declared that the bee was the most important creature on earth. Ninety percent of bees have disappeared in recent years. One of the most interesting factors that come into play here is how unconcerned most people seem to be. The ramifications if the bee goes extinct are so large that we don't fully comprehend them, and you much of the world seems to have buried its head in the sand to this fact. A 1998 estimate suggests that every year, 200 billion dollars worth of crops are pollinated by bees.

Between 2007 and 2013, 10 million colonies were lost, twice the amount we would expect to lose under normal circumstances and the most unprecedented loss in the history of beekeeping. It is easy to see how loses at that rate could quickly become unsustainable.

Theories as to what the reason is for this sudden demise in the numbers of honeybees are varied, but though we may debate the causes, it is clear that more needs to be done to protect them. Scientists, professional beekeepers, and agricultural officers are all attempting to address this issue, but surprisingly, much of the effort has also been applied by small home beekeepers. Backyard beekeeping has seen a massive surge in numbers ever since the news of the colony loses started to become widely publicized. That may turn out to be important because these small home operations may turn into an important reserve should the downward trend continue. Now instead of the majority of bees being kept in close proximity to one another by industrial-sized beekeeping businesses, there are thousands of small hives more widely spread, and this may help save the bee if they are dying from some form of transmittable disease.

What tends to happen with colony collapse disorder is that the worker bees simply disappear. The keepers open their hives to find the queen still present, along with plenty of food and just a few nurse bees, but the remainder of the colony has vanished. One of the problems is that bees are tiny and so millions of them could die, and nobody would be any the wiser as to where or why. Worker bees travel long distances to source food, and so even large numbers of individual insects would not be noticeable when scattered over that sort of geographical range.

Several theories dominate the suspected causes list for CCD. Varroa and acarapis mites maybe two while immunodeficiency's is another that is commonly suggested. Global weather change could be the cause of some of the problems, but it seems unlikely to account for such a large loss of colonies. In the US especially, but also in other countries now, bees are used primarily for pollination. This requires that they often are transported for long distances. They may be trucked to one area to service a flowering crop there and then moved again to do the same thing on a later flowering crop after that. This may take place several times a year. The transport and the relocation at each new destination are stressful, and it could have been the portal for some yet to be discovered disease or health problem. Though this idea is not outside the realms of possibility, what we do know is that colonies that are not subject to this moving are also being lost. It could be that the transported bees have developed a disease that they subsequently transmit to other colonies, but if that is the case, this disease has yet to be pinpointed.

Probably the potential culprit that has drawn the most attention is the pesticide family called neonicotinoids. First discovered by the Shell Corporation was taken over by the pharmaceutical company Bayer although many other companies now produce neonicotinoids based insecticides. Soon after this line of pesticides was first offered to the public in 2008, it took over twenty-four percent of the global market. Between 1990 and 2008, sales rose from $155 million per annum to $957 million per annum. In the US, it was used as a seed coating on 95 percent of corn and canola crops and most of the fruit and vegetable crops in commercial production. In short, neonics, as they are commonly known, were virtually everywhere in the agricultural crop production world at a very similar time to which we started to see the massive drop off in bee numbers.

This has become a very contentious issue in the scientific world, with representatives from both camps pointing fingers at one another. The environmental lobby remains convinced that they have found the cause while the big pharmaceuticals are adamant that there is insufficient scientific evidence to support their claims. In 2018 EU countries posed a total ban on neonics, with most government bodies agreeing that they posed a negative impact on bees. The US did not feel obliged to err on the side of caution. The Obama government did ban use in farms within National Wildlife Refuges, but that was subsequently overturned by the incoming Trump administration.

Representatives John Congers and Blumenauer presented a bill entitled the Save American Pollinators Act to the House of Representatives in 2013. It was subsequently sent to the congressional committee from where it never left.

For those of us who are not scientists, the whole argument is probably too technical to form an opinion truly. Both sides of the debate have powerful lobbying capacity, and there is a ton of money involved, as well as jobs and export markets. It would seem that the European decision to follow a cautious approach until more evidence can be gathered is a wise one, but that is clearly not an opinion that has prevailed in the United States.

For the would-be beekeeper deciding whether or not to invest time and money into a hive or two, the above information might be somewhat off-putting. To balance that, there is the knowledge that now, more than at any time in history, your efforts could be crucial. If you are away from anywhere that neonics are being used, such as in a town or city, it could be that you never are affected by CCD. If that were the case and it was replicated in many areas, it would provide more evidence that pesticide use could be at the root of the problem.

Regardless of whether or not chemical use is causing an impact, beekeeping is something that will always pose a challenge. Hives need to be kept free from mites and disease and proofed against rodents and other invaders. These issues all need to be considered, but they need to be weighed up against the rewards. The threats are visible and well-publicized. The benefits are far less tangible and will certainly always gain less attention in a world where the press seems to thrive on the negative rather than the positive.

There is a learning curve when you first start getting to grips with the whole beekeeping process, and it can appear quite daunting. I really urge you to press on. This is a natural process, and as you begin to engage with it, you will find that things start to flow with a natural rhythm. Don't imagine you will learn everything that there is to be learned but pace yourself and evolve rather than trying to absorb everything at once as though studying for an exam. People have been studying bees for millennia, and today, though they remain by far the most studies insects on the planet, people are still learning about them. What is more, we seem to be learning from them.

BEEKEEPING

Where studying for most subjects can become a chore, beekeeping is so interesting that the information you learn will be absorbed almost by osmosis. Regaining touch with nature requires just one key issue like this to set off a whole chain of learning from the natural world. Part of being a good beekeeper means that you will learn about the pests that threaten your hives, and the different flowering plants that sustain them. As your bee husbandry improves, so will your respect for the environment that they live in. Flowering plants and fruit trees in the vicinity will bear more abundantly, and this, in turn, will provide habitat for all sorts of other creatures. You will, I am sure, develop a strong desire to move away from pesticide use and will probably also want to stop using herbicides too. There are natural alternatives to these products, and soon, you will find yourself wanting to study more about these as well, and so the whole learning circle increases and all just because you installed one little beehive.

Unless total solitude is your thing, beekeeping exposes you to a group of likeminded people. At first, they will be your guides, but if you stick with this project, it won't be long before you find yourself mentoring others on their journey. A brief word of warning here; beekeepers tend to become so passionate about their subject that they want to evangelize it to everyone they interact with.

BEEKEEPING

Bees are not expensive to keep, and any capital outlay can quickly be recouped by selling the honey you produce even if it is just to close friends. Most of the honey sold in supermarkets is adulterated in some way. The fact that you are producing pure honey and that you have a pretty good idea what the main flowers that went into making it was, means that you will soon find a steady demand for anything you don't consume yourself.

All kids are fascinated by what goes on in that stack of boxes you keep in the back yard. You might believe you are the most boring parent or grandparent in the world, just step outside in that baggy white suit with its ridiculous hat, and you will have them clambering to join you. Unless Superman himself shows up, you will undoubtedly, suddenly be the most interesting adult they know. Personally, though I can't prove it, I think Superman has got nothing on a half-decent beekeeper.

Kids really quickly overcome their fear of being surrounded by hundreds of buzzing insects, and they then make dedicated students. Without any real effort on your part, you will be helping to establish the next generation of environmentally savvy kids who will hopefully go on making the planet a better place than our generation has done.

So just to recap a little, beekeeping offers you a chance to improve the environment and offset some of the damage we have all done to it. It puts you on a fascinating learning curve and helps you produce better flowers, fruit, and vegetables. If you want to, it can expand your social circle and even help you get on with your kids or grandkids while at the same time providing you with a wide variety of useful or even saleable produce. On top of all that, beekeeping can be one of the most relaxing things a person can engage in without actually having to engage the services of a full-time masseur.

For those who become so addicted that they want to expand their beekeeping beyond a mere hobby, there is a range of career paths open to them. The obvious one is to expand the number of hives you own and become a full-time apiculturist, but that is by no means the only option. There are opportunities in research for those with a strong leaning towards the science of the industry. For those who prefer to remain hands-on, there is often a demand for beekeepers on a seasonal basis with the bigger operations. Some keepers spend the summers in different hemispheres working on bee farms in the US and then flying down to Australia or New Zealand to work the hives in the southern half of the planet. There is also a wide range of satellite businesses, such as teaching, speaking events, and writing about beekeeping. Although not officially recognized in the US, there are people who specialize in making therapeutic medicines from the many products that each hive generates.

I know that I have painted a blissful picture by focusing so heavily on the benefits of beekeeping. In the interests of fairness, I need now also to explain some of the many disadvantages. Just as good farmers form a relationship with the animals they are stewards of, beekeepers develop a relationship with the insects that they look after. You will know which hives aggressive, which ones are preparing to swarm and which ones aren't thriving as well as they should be. Eventually, such knowledge will become almost intuitive. When something goes wrong, it can be really depressing, and there will be days when you come home from inspecting your hives and swear that you are never going to keep bees again. The threats to bees can seem overwhelming on occasions. The hive risks attack from mammals, insects, viruses, mites, birds, and even fungi. On top of all that, there is this lingering threat from pesticides that we don't fully understand and may never get to the bottom of. You will learn more about the non-chemical obstacles as you progress through this book. As far as the chemical threat is concerned, all beekeepers are still awaiting the outcome of those investigations. For now, all we can do on that front is trying to ensure our bees are kept away from possibly harmful, manmade threats, while ensuring that we do all in our power to make a life for them as near to perfect as we can.

BEEKEEPING

CHAPTER 3

TYPES OF HONEYBEES

There are over twenty thousand different types of bee, and yet only seven of those are honeybees. From time to time, when you discuss the decline in bees that we are currently witnessing, someone will tell you that bee loss is overdramatized and that there are plenty of other pollinators that can fill any gap that may be made by a decline in their numbers. He may even quote the figure above and tell you that honeybees make up just a fraction of the overall bee population. These people don't really grasp just how much of the pollination process is carried out by the humble honeybee and how difficult it would be to replace them.

This is not to suggest that people are not looking at exactly this solution to our looming pollination problem. In the US, scientists have long recognized the threat from CCD and begun trials with other bees. One bee called the blue orchard (Osmia lignaria) had shown some potential. Just 200 hundred of them can do the same amount of work as 100 000 honeybees, but here is the catch; they are solitary bees that lay their eggs in holes bored into trees by other insects. On an annual basis, even under pristine controlled conditions, the blue orchard can only increase its numbers by a factor of three to eight. A single queen honeybee with a few workers can develop a colony of 20 000 bees in a matter of months. That colony system in which the queen plays the role of a very well attended egg-producing machine is difficult to compete with. What is more, the blue orchard produces no honey.

Honeybees are far from the only pollinator, which is why only about one-third of crops are pollinated by them. Many other insects perform a similar role including, flies, wasps, butterflies, and bumblebees. In fact, with its large hairy body, the bumblebee is a more effective pollinator. There are two problems that come into play here. One is that the honeybee is the one insect we have been able to farm on a large scale, and thus, we can place them where they are needed, and the other is that for honey production, no other insect even comes close to that of the honeybee's production capacity.

We could probably learn to live without honey and just let other insects fill the gap if the honeybee numbers declined to such an extent that they were no longer viable to farm. There is another hitch in this thinking, however. Nearly all insects have seen dramatic declines in recent years. Again, reasons vary wildly, and with many of these insects, there has been far less research than with the honeybee. Habit loss is certainly a large part of the problem, and if insecticides are killing honeybees, it will definitely be killing many other insects as well. A recent study published in Germany revealed that within the last thirty years, there had been a seventy-five percent drop in the number of flying insects. The study was performed in three areas under natural protection, where it was assumed insects would be at their most abundant. All of this just goes to drive home the point that the need for careful protection of the honeybee is crucial.

This is where you come back into the picture. Hopefully, by now, you have become convinced of the importance of beekeeping and are still considering pressing forward with expanding your understanding and knowledge of the subject. Earlier in the book, we mentioned that there was only one species of honeybee compared to twenty-odd thousand other bee species. What we are going to look at now are the different subspecies of the honeybee in order to decide which one is most appropriate to your situation.

There are twenty-six strains of the honeybee, but only seven that should really concern your choice when creating your beekeeping stock. Each has different characteristics that will influence your decision. You should also speak to local beekeepers in your area as they will know of factors that come into effect, such as the bee's ability to tolerate local conditions and which of the subspecies thrive best on local flora.

The honeybee was first brought to the US from Europe in 1622. There is fossil evidence of a local honeybee having existed millions of years before that, but it subsequently went extinct. Native Americans are said to have referred to honeybees as 'white man's flies.'

The Italian honeybee (Apis mellifera ligustica) is a very popular bee for honey production. They originated in the south of Italy but have been widely spread through the activities of human beekeepers. They are a great general purpose bee and are highly favored by commercial beekeepers. There predominantly yellow coloring makes them quite easy to identify.

They are prolific honey producers, but at the same time, they consume a lot, so they need to be well-stocked before the winter. There is also some evidence that they are more prone to disease than some of the other subspecies. They are not inclined to swarm terribly easily, which is a useful trait as you don't want to lose your stock. They are also renowned as thieves and will readily steal from hives of weaker subspecies.

The Carniolan bee (Apis mellifera carnolia) is believed to have originated in Slovenia and the Austrian Alps. They are the second most popular bee to farm, and as the Italian bee, they are ideal for the first time beekeeper as well as the larger producer. They are favored for their good nature and the fact they will not sting unless really provoked. They are a duskier brown color than their Italian cousins. They also have a long tongue, which is an advantage when foraging. Because they originated under fairly adverse conditions, they are good at overwintering, and they form tight clusters in the hive during the colder months and use little resources. Another of their attributes is their ability to cope with insect invasion. The main downside of this subspecies is that they swarm quite readily. It is thought that because they don't like to be overcrowded, so the beekeeper needs to ensure that they have plenty of space in the hive.

Caucasian bee (Apis mellifera caucasica) - This bee, once very popular, has fallen out of fashion, especially with commercial producers, because it does not produce vast amounts of honey. Other than that, it bears many of the same attributes as the Carniolan bee. One additional attribute is that they are very neat producers and almost insist on filling a frame before moving onto the next one. This means that they are well-stocked in winter, and their overwintering food source is both closes at hand, and they are kept warm.

If you live in an area where harsh winters are the norm, then this is a bee you might want to consider. Amongst traditionalists, they are going through a bit of a revival, and they are often crossed with carnelians to increase honey yield.

The Russian bee - Imported as recently as 1997, this bee hails from the east of Russia. One hundred queens were brought into the US by the Department of Agriculture in order to test them as a method of overcoming Verroa and other mites. The bees are certainly much more resistant to these pests, but other problems have emerged. They are quite aggressive and cross breed readily with some of the varieties that have been here much longer. It has been noted that when they do that, their resistance levels go down dramatically, thus reducing the primary reason for having them. Much is still to be learned about them. For example, most colonies will produce a new queen cell when they are ready to swarm. Russian bees seem to always have a queen cell on standby. There are still controls on who can breed them, and this is not a variety that should be considered by the debutant beekeeper.

The German bee (Apis mellifera mellifera) - This bee originated in Russia and Northern Europe. It is actually endangered in Germany but still used widely for honey production in many European countries. It was one of the early imports to the US, and its genes are thought to be found in many of the colonies there. It is easily recognized for its dark, almost black coloring. They are very winter tolerant but have not fared at all well in the US due to their exaggerated intolerance of certain diseases such as both the American and European foulbrood. There are still a few real enthusiasts who are experimenting with disease tolerance in this bee, but it is not recommended for the beginner.

The Buckfast Bee - This bee was bred at the Buckfast Abbey in the United Kingdom in the 20th century to try to overcome the threat posed by tracheal mites. The monk, Brother Adam, who was head of beekeeping at the Abbey in 1919, noticed that these mites were causing widespread losses, and so he set about crossbreeding with any hives he could find than seemed free from of the mites. He crisscrossed Europe to develop his stocks, and the bee that eventually resulted from his efforts was named after the Abbey. These bees are hugely popular in Europe, where they have a reputation for being both gentle and reliable honey producers. In the US, these bees are slightly harder to come by, in part because they do not swarm regularly, and so stock does not increase. They are known to crossbreed easily, and that often results in loss of the traits for which they were originally bred.

The Africanized bee - This bee is mentioned more because of the importance of understanding them rather than suggesting them as a potential breed for stock. In fact, they are not African bees at all. They originated in South America, where they were part of a scientific program. As is usual, these bees were being bred primarily to look for resistance to mites and other parasites and also for increased yields. In 1956 twenty-six of the hives managed to escape, and they gradually made their way up from South America to the US amidst much hype and scared mongering, which included the production of several killer bee films. The bees are aggressive but pose nowhere near the threat their reputation would suggest. Some beekeepers are trying to work with them, but to date, their results are not very promising.

In South America, these bees are one of the most widely farmed bees, but that is because they have crossbred so widely with other domestic stock. They have more guard bees per hive and are more inclined to act defensively

Now that you understand a little about what options are available to you, here are some of the factors you need to bear in mind when making your decision. Firstly, what experience are you bringing to the table? I am guessing that you are pretty new to this; otherwise, you would probably have already worked out which bees are best for you.

For those thinking of dipping their toe in the water, it is always best to start off with more gentle bees. Production is not that much lower, and they make things so much easier. We've discussed kids and neighbors, and the last thing you want is someone getting stung. As your experience develops, you will become so accustomed to managing your hives that you will get good at predicting what will alarm your bees or trigger defensive behavior. To begin with, just work with stock that is easy to handle. Most beekeepers never really bother to move onto more difficult to manage bees because they just don't see any reason to. If you are a commercial beekeeper and need to maximize production, or you decide to delve deeply into breeding for specific conditions, then things are different. These people almost always get to that stage after years of experience.

The experience of other local beekeepers really cannot be overstated here. They will know more about local bees and local conditions than you will be able to learn from any book. They will be able to advise you about the prevalence of both disease and parasites and will have already worked out what bees are best suited to those adversities. They will also know if there are farmers in the area who have a predilection for spraying and at what periods they are most likely to spray. That is valuable knowledge because you should try to keep your bees housed on those days.

Much of what you have just read about types of bee will guide your thinking, but it is likely that nobody knows for sure exactly what variety of local bee they are farming. The bees are likely to have crossbred to a certain extent, and the beekeepers will just be happy that their stock is thriving rather than carrying with it a high degree of pedigree.

One factor to bear in mind when taking advice is that you should find out what motivates the person you are listening to. Mainly, smaller beekeepers will have a similar ethos to your own, and their priority will be stewardship first, profit second. With some commercial operations, this ethos is reversed, and the advice you get will differ. First of all, you need to examine your own goals and motivations closely, and then you need to see if your mentor's correspond to those. I am happy to say that I have met very few beekeepers, both professional and otherwise, who did not put a great deal of emphasis on the well being of their stock.

CHAPTER 4

TYPES OF HIVES

So you've decided that beekeeping is for you and you have made the decision that you want to get some bees. Where does one go after that? The first thing to do is decide on is what type of hive you are going to use. Here again, the range of options is wide, and you will need to expand your knowledge again. You have two main priorities to consider here. What hive will be most comfortable for the bees and what hive will be most practical for you to work with. We have already seen that the very first farmed bees lived in hives made from clay. We all also probably recognize those old fashioned hives that were woven out of basket material and which often decorate honey pots still today. They were called skeps, and when the beekeeper wanted the honey at the end of summer, he just chased away the bees using smoke. After the honey was removed, the skeps were simply burned. These bee houses might not be in use much today, but it does serve to demonstrate that bees are fairly flexible in where they will take up residence. Most beekeepers will be able to regale yours with stories about being called out to remove hives built up chimneys, inside hollow walls, or in cracks in trees.

While the bees may be fairly undiscerning, you need a hive that gives you easy access and the bees the ability to expand and make plenty of honey. The more comfortable the hive, the more likely the bees are to remain and to flourish.

Hive Components

You are about to see that there is a wide variety of beehives, and therefore, not all of the components listed below are relevant to the entire hive. The terminology is fairly standard, and so it will probably be helpful if you understand it now as it will have an overall pertinence.

The hive stand is at the very bottom. It keeps the hive off the ground and often incorporates a landing board from which the bees can take off and alight. Not everyone bothers with this stand as its primary role is to keep the hive clear of the ground, and a couple of cement blocks will do the job just as efficiently.

The bottom board is a flat wooden board or a wooden frame screen. These days the screen base is becoming more widely used as it allows ventilation but particularly because it lets pests such as verroa mite fall through rather than remain within the confines of the cage.

The deep hive box - This is the bottom box in which the breeding takes place and where food is stored. To make life more confusing, a variety of terms are used for this compartment. You will hear it called the deep, the brood box, or simply the box. This is the most important part of the hive. It is where the breeding takes place and where the queen lives. In some areas, the second deep is added on top of the first one. The bottom one plays its normal role, and the higher one becomes more of a larder area and is particularly useful in areas that experience long hard winters where extra reserves are important.

BEEKEEPING

The queen excluder - This is an essential part of the hive. Because you don't want the bee to enter the main honey area and laying eggs in cells where your honey comes from, she needs to be kept down in the brood box. The simple solution is to place a grill between the two through which the workers can pass to perform their duties but through which the larger queen does not fit. This may seem a fairly rudimentary solution, but it was the discovery of what size grill was appropriate that enabled the advances in modern beekeeping that we currently see.

The super - This looks a lot like the deep box but is considerably shallower. It is filled with frames for the honey to be drawn out in and sits right on top of the queen excluder. As it begins to get filled with honey, another super can simply be placed on top to expand the bee's storage space. The honey in the supers is what you can harvest. What there is in the deep is the bees supply, and it must be left for them. Supers come in two sizes; shallow and medium. The only difference technically is that one can hold more honey than the other. The choice you will need to consider is the weight you want to be handling. A shallow weighs about forty pounds when full of honey, while a medium can weigh up to sixty pounds.

There is another option here. You can use mediums throughout the hive and just double them up to form the deep or brood chamber. The bees will not mind, and it means that you only need one size of the frame right through your hive. This can be handy in terms of mass production if you are making your own frames.

Frames: These hang in the hive, and it is on these that the bees suspend their comb, produce brood and store honey. When fitting them, you can pre-fit them with manmade wax foundation strips. These look very much like sheets of pasta, and the reason you would fit them is it saves the bees from having to make wax for the comb. Instead, they just draw out the foundation strip and use that, thus saving them energy and allowing them to focus on honey production. They are not critical, and in some of the hives, we will be looking at frames that are replaced by a simple wooden hanging bar from which the bees suspend their own comb. Frames are set exactly six millimeters apart, and this allows the bees just enough room to perform their chores but not enough to get too cold.

Inner cover: This has the appearance of a flat tray with a hole in it. The hole allows ventilation, and some people are now switching to inner grills rather than sold covers. This will obviously improve ventilation but will make it harder to keep the hive warm. That is a decision you should make based on local weather conditions, but again, this is something where local beekeeping knowledge could come in handy.

Outer cover: the outer cover or lid is what keeps the hive watertight, and it just sits on top of the hive with an overlapping lip all round. It will obviously need to be waterproof, so it should be covered in roofing aluminum or some other waterproof material. They can blow off in extreme weather, so weight them down if you expect to experience heavy wind.

There are one or two other bits and pieces that will make your life easier. Over winter, it is handy to have a sliding cover so that you can reduce the size of the entrance to a minimum to help retain heat. Mouse guards are also fitted over the winter months as the hive is both warm and full of food that could tempt rodents to move in. In the summer months, when the bees are active and have guards stationed, this does not become an issue.

The Langstroth Hive

Patented in 1852 and first produced in 1920, the Langstroth looks a little like a narrow set of drawers. Its inventor was the Reverend Lorenzo Lorraine Langstroth, and today it, or derivatives of it, is the most commonly used hive and the one most of us are probably most familiar with. The clergy has a long history of beekeeping and was instrumental in many of its formative advances. Langstroth was no exception, and his hive certainly has withstood the test of time.

The hive itself is divided into compartments, which gives it its draw like appearance. The bottom and largest section contain the nursery section, while the higher compartment contains the honeycomb and the honey. As the comb in the higher sections is filled, then more compartments can be added. The hive can be opened, and the honey removed without causing any undue stress to the lower nursery.

There is normally a flat roof, which consists of an inner roof for insulation and an outer roof for weatherproofing.

The Low or Horizontal Hive

In principle, this hive is very similar to the Langstroth but in a horizontal format. The bees will naturally station the brood section nearest the entrance and then start their honeycomb operations after that. In some ways, it is easier to use, especially for the disabled or for children. It normally comes with a small window like viewing ports to enable checking of the hive over winter without opening the hive.

These hives are a variation of the Langstroth, and as such many of the Langstroth fittings such as frames can be used. The reason that they don't seem to have become as popular as their vertical ancestors are that they are more expensive to make and don't transport as easily

The Warré Hive

Named after its inventor Abbé Warre was another clerical beekeeper, this time-based in France in the early 20th century. Langstroth used some of the Abbe's ideas to develop his hives. Abbé wanted to build a hive that was cheap, easy to use, and which worked with the bee's natural instincts rather than against them. The hive consists of a series of boxes, but this time the brood box is at the top, and the honey is made below that, which is a more natural approach for the bees. Instead of frames, the boxes contain bars on which the bees suspend the comb much as they would in the wild.

The roof is sloped to shed water and contains a space known as a quilting box in which insulation such as straw can be placed to keep the hive warm in winter. The straw and roof design combine to allow movement of air in the hive.

The advantage of this system is that the hive needs only to be checked about twice a year, so disturbance is kept to a minimum. When the honey is harvested, it is taken from the bottom, so not disrupting the nursery. Windows are often added to the boxes to make checking easier. This system is often preferred by people who want their bees to lead the most natural life possible. The queen is not excluded from the honey boxes, and the harvested comb will have contained brood. The queen goes onto a new comb to lay the next eggs. This means that all eggs are laid into a fresh and disease-free comb.

The disadvantage is that the bees will stick the boxes together firmly, and some careful lifting will need to take place to separate them. Fortunately, component parts are cheap, so if the damage is done, it is not too expensive to replace the damaged pieces.

The Top Bar Hive

Sometimes referred to as the Kenyan or Tanzanian hive, these are really simple hives that have drawn inspiration from both the Warré and the long hives. Like the long hive, the bees are housed in a horizontal box. The first noticeable difference is that it is trapezoid in shape rather than rectangular. Like the Warré, there are no frames, and the bees simply suspend the comb from bars at the hung across the top of the box. When the beekeeper wishes to harvest the honey, he simply opens the hive, removes the bars individually, and cuts off the comb. The whole of the bow lid is hinged so that it can be opened easily, and there is space beneath the lid to allow for insulation.

The bees enter the hive at the bottom of one of the face panels, and a very simple plank can be suspended in the hive to reduce or expand the interior size according to the size of the colony. This means that reducing the interior space over winter is easy, and the bees are kept warm and comfortable. Because of the shape of the hive, the suspended comb comes out in an almost triangular shape, which means it is less heavy and less likely to snap off when being lifted from the hive for harvesting honey. This suspended comb system is very much in keeping with how the bee would make comb in the wild.

The Flow Hive

Beekeepers are always looking for systems and procedures that will make their lives easier. This is particularly true for professional beekeepers that must manage hundreds and sometimes thousands of hives. The flow hive is one of the most recent innovations in the beekeeping world and is highly controversial.

BEEKEEPING

People either love it or hate it, and there seems to be no middle ground.

The flow hive works in conjunction with the Langstroth hive. The bees use the same brood box, but on top of that is placed the new system, which essentially consists of a series of plastic honeycombs. The idea is that the bees will make the honey in these combs that you the beekeeper have provided in the normal way. When they are full, the beekeeper inserts a long key, which, when turned, crushes the individual cells causing the honey to start to run out. It falls into a trough and then runs out through a tube. The beekeeper then simply needs to place his jars beneath it and catch the flow of honey.

There are advantages and disadvantages to any new system. In this case, it is a new and technical system that has been introduced to a very old and traditional one. Some of the advantages are that it is not necessary to open the hive as often, and therefore there are fewer disturbances to the bees. It is far more convenient and less physically demanding for the beekeeper.

The downsides are that the bees are making their honey in plastic instead of their natural wax cells. The hives are far more technical than any of the more traditional ones and are therefore not cheap. If you are harvesting thick honey, which we will deal with at a later stage, the honey will not flow. If not all of the cells have been capped, there will be a number of bees still working on them, and they are likely to be crushed. The only way to assess what percentage of the cells has been capped is to open the hive, and that removes one of the advantages we have mentioned.

These hives have been heavily promoted among people that had never kept bees before, and one of the conclusions these people reached was that they could get into beekeeping without very much of the effort. A simple twist of a key once in a while would be all that was required to have great home produced honey made right on your doorstep. That is not the case, and proper bee husbandry is still required. The flow system brings with it some advantages, but don't be fooled into thinking that you can ever avoid the important issue of caring for your bees. If you are going to profit from their labor by stealing their honey, the very least that is required of you is that you should care for them in the best possible manner.

We have now had a look at the main types of hive. There are plenty of others out there, but when you understand these, you will pretty much know how most of the other options work. Other than the flow hive, there have been years and years of work done with these hives, and whichever route you choose to go down is going to have been well tried and tested.

It now remains for you to decide where you are going to acquire your first hive. There are many companies that specialize in nothing other than beekeeping equipment, and a visit to one of their stores is an education in itself. They will be able to guide you further, and you will be able to see firsthand how the hives work and the quality of the build.

Like anything else in the world that you choose to buy, the internet now provides you with an easy buying option. Sometimes these hives come with an element of self-assembly involved, so look out for that possibility. Always check the comments left by previous buyers, and if there are none or very few, then you might want to look around a little more. Internet suppliers are often cheaper even with the shipping involved, but unlike purchasing a book or a mass-produced product, you have no way of feeling the quality before you become the new owner.

I know that I have promoted the local beekeeping club quite heavily, but that is because I really believe that the fee knowledge to be gathered at these little groups is simply invaluable. Almost all of them have some sort of arrangement whereby you can purchase equipment, including hives, at a slightly reduced price. There are also so often people in these clubs who have extra bits and pieces that you may well get a second-hand hive for nothing or next to nothing.

The final option is to build your own hive. For anyone with slightly more than basic carpentry skills and a good set of tools, building a beehive is not a big deal. One of my friends is now a professional beekeeper with well over two hundred hives, all of which, he built himself using old pallet wood. There are plenty of plans on the internet, so even if you can't borrow one to copy, you can still create your own. One word of warning here, if you are going to make a hive, first ensure that any wood you use has not been treated with any sort of chemical. Treated wood is particularly bad for bees that live in constant proximity to it.

Where to Place your First Hive

You can make or buy the best hive in the world, if you don't position it correctly, then the bees are either going to leave, or they are not going to thrive. Location is probably one of the things that you considered when buying your home. Well, the bee location is equally important.

In the northern hemisphere, it is preferable to have your hive face south. This will mean the rear of the hive is most likely to get hit by the worst of the cold winter winds. If the hive catches the early morning sun, the bees will warm up more quickly, and this will get them foraging earlier. They can't fly until they have warmed up. If the back of the hive is protected by a hedge or fence, that will make conditions in the hive warmer. The bees will be coming and going from the front of the hive, and so this flight path needs to be reasonably clear. You don't want the workers having to avoid obstacles or pedestrians as they approach or take off. If the hive is in at least partial shade, it will prevent the bees from overheating during hot summers. Bees cool the hives by placing workers in the entrance to flap their wings and move the air. If they don't need to do this, then that energy can be better used elsewhere. Also, hot bees can become angry bees.

Bear in mind that though it will not be heavy when you place it, the hive is going to become considerably heavier as it fills with bees and honey. Make sure that it is staged on a level position that is not going to subside.

BEEKEEPING

Personal Equipment

There is plenty of equipment that you could by when you first start keeping bees. I think that it is better to start with the basics and then expand from there than it is to buy everything people will try to sell you, only to find it is really not useful.

You will need a bee suit, to begin with. Many beekeepers stop using a suit as they become more confident, and you will probably have seen these people and marveled at the way in which they confidently move through clouds of bees without any protective gear. Well, the truth of the matter is that bees have different moods, and even the hardiest keepers will probably have occasions when they decide it is wise to don a suit. For the newbie, the suit is really essential. You will be embarking on a sharp learning curve, and you don't want that disrupted by having to deal with stings. As you become more confident, you will very likely start working with just a veil, but there will still be times you need to turn back to the bee suit so you may as well save yourself the pain and purchase one straight away.

Veils: These are simply mosquito netting type material attached to some form of headdress. You can save money and make one yourself, or you can purchase one. The main thing is that it prevents the bees from getting to your face. Make sure that when you are wearing it, the bees don't like the veil and then drop down your shirt as this is extremely disconcerting. I tend to favor the pith helmet type as it keeps the veil clear of my face, and it can tuck it into the hat itself when I don't need it.

The smoker: This is the metal coffee pot type affair that has a bellows attached. It is fed with a variety of different materials to create smoke, and different beekeepers tend to have their own ingredients for fuelling them. These range from straw to old pieces of hessian sacking. I have even known keepers who use dried horse dung. The larger the fuel chamber, the better is the important thing when choosing a smoker. You don't want them going out as this causes all sorts of problems once you have the hive open. It is worth doing a few dummy runs and getting to grips with both your smoker and the fuel you will be using.

Gloves: To start with, these are indispensable, but they will also probably be the equipment you most quickly stop using. Gloves that are thick enough to prevent sting altogether make you lose dexterity, and working in a hive, especially one well sealed with propolis, can be quite fiddly. Some people try to get the best of both worlds by wearing thinner gloves, but you need to accept that these are not entirely sting proof. Gloves need to be cleaned from time to time. Each sting leaves a little scent that is designed to guide other bees toward the target, so if they smell that, they will zone in on the gloves again and again.

Hive tools: These are pretty much vital. You will mainly use then for levering off top panels and loosening frames from the grip of the propolis coating that the bees will apply to them. Hive tools come in two sorts. The flat tool and the J shaped. The J shaped is to use for levering up frames, but actually, they are both pretty similar when it comes down to practicality. One useful suggestion is to pain whichever one you opt for in bright reflective paint because they have a nasty habit of getting lost in the grass just when you need them.

CHAPTER 5

THE ROLE OF HONEYBEES

Before we jump ahead to that point, let's just take a look at the types of bees that make up a colony so that you know exactly who does what. You probably already have an idea of the three main types of bees. The workers are the guys who do all of the days to day stuff. The drones are the male sperm carriers who don't do very much at all, other than providing sperm, and the queen is the top of the pyramid. She is fed, cleaned, and massaged by her workers so that she can focus on her primary role in life, which is simply to lay eggs. We all learned that in junior school, so we have a basic idea of how things work.

Actually, things are more complex than that, and that is what makes the inner working so of your hive so fascinating. We have been studying bees for hundreds of years, so we have a pretty good idea of what is going on. Despite all this academic and scientific attention, there are still things about bees that we just don't understand.

BEEKEEPING

One of the most important things to get into your head is that bees are not individual units. They are more like a highly complex organism. No part of the hive can survive for any length of time without the other. Lose the workers, and the queen will die. Lose the queen, and the whole thing just falls apart. The drones are just men and, as such, are pretty dispensable, but they do play an important role, nonetheless.

An average healthy hive will contain approximately 80,000 bees at the height of summer. The vast majority will be female worker bees. There will only be a single queen who is the mother of every one of the worker bees and is also the only fully sexually developed female in the hive. The queen is longer than the workers with a thinner abdomen. In theory, this makes her easy to spot, but the difference is not that great, and finding her amongst 80 000 other bees is a bit of an acquired knack. To make things a little easier, the beekeeper will often mark the queen with a tiny dot of the special waterproof marker. For a newbie, spotting even a marked queen is not always simple when you are standing amidst a cloud of flying bees wondering if you really haven't got any allergies that your mother didn't warn you about. Don't worry; things will get easier.

Marking the queen is another trick that takes a bit of practice. This is where having an old hand to guide you starts to really pay off. For starters, you need to spot her; then, you trap her with a special clip. She is then gently coaxed (prodded) into a marking tube, and the marking pen is applied to the back of her abdomen, leaving a dot. A few seconds are needed for it to dry before she can be placed back in the hive, where she will be none the worse for her adventure. Your hands will be shaking, and you will be sweating lightly, but don't worry about that. The next time will be a cinch.

To make life easier down the road, there is a color code for the dots. Blue is for years ending in 0 and 5, white for years ending in 1 and 6, yellow for years ending in 2 and 7, red for 3 and 8 and finally green for years ending in 4 and 9. This universal system is used so that you know the age of your queen. When looking for the queen, start by looking at where new eggs are being laid. As this is her primary purpose in life, it is the most likely place for her to be. Another clue as to her whereabouts is the behavior of the worker bees. They constantly attend to her, so there is often movement in her direction, and she always has a small crowd of servants in her immediate vicinity. As far as marking her is concerned, you can practice your technique by doing some test runs on the drones.

Drones are also a different shape to the mass of workers. Like the queen, they are larger, but the back of their abdomen is more rounded. This is because they don't have a sting, which is another reason they are good subjects to practice on. A working hive will produce about two hundred drones per year. The workers build larger cells for drone eggs, and the drone will only have half the chromosomes of a worker bee. They contribute nothing to the hive. Instead, when the world warms up enough for their liking each day, the drones will fly out and sit around together, often in a high tree. There they will shoot the breeze, discuss the latest sporting events, and keep an eye out for any passing virgin queen.

These are rare events. When one is spotted flying past the drone assembly area, feigning nonchalance, she is immediately pursued by all the drones. The swiftest will reach her first, and they will mate on the wing. In the process, the drone's genitals will be torn out, and he will quickly die. Even if the drone does not get to mate and therefore does not die in a blaze of short-lived glory, his future is bleak. The workers won't tolerate a non-producing member over the winter, and so the drones are pushed out of the hive, where they will quickly die. The queen collects the semen in an internal sack. She may mate with up to twenty drones on a flight and can make two or three flights before the sperm holding sack is full. Once that happens, she will be carrying enough sperm to last her the remainder of her life. A queen can lay up to 2000 eggs per day, which are more than her own body weight in eggs, so it becomes more obvious why she needs to be so heavily pampered by her workers.

On her return from each of these flights, she will immediately be pampered and fed by the worker bees. Her value to the colony is becoming critical. If they lose a queen late in the year, they will not be likely to create another one before winter sets in, and the whole hive might die out. Mating flights require a minimum temperature of 16 degrees Celsius. For the beekeeper, the queen is equally important, and after she has completed her mating, some will clip one of her wings to prevent her from flying off again and possibly becoming lost.

It is the workers who decide when to start preparing a new queen. They do this when the older queen starts giving off fewer pheromones, and they become aware that she is about to become less productive. The queen to be can be created from any of the newly laid eggs, but once the workers decide that a queen is required, they will start feeding what is known as royal jelly to the newly hatched larva. If the original queen is lost for any reason, the bees will immediately start trying to produce a new queen. The diet of royal jelly is what turns an ordinary bee into a queen. It is produced from a glad in the head of the worker bee and is much richer than the food that normal eggs get given.

BEEKEEPING

Understanding the role and life of the queen is one of the most important jobs for any beekeeper. The bee dictates the life of the hive, and so the beekeeper may decide to replace her for any number of reasons. It may be that the colony is very aggressive, in which case just introducing a gentler queen will have a calming effect on all of the bees in the hive. He may want to introduce some other trait by crossbreeding with bees from different sub-species, or she may simply be getting too old. Queens normally live for two or three years.

All of these are reasons the beekeeper may have, but the worker bees themselves are also able to decide to change queens. The queen produces pheromones, which are known as substance. It is these pheromones that give a signal to all of the bees of the colony as to the queens well being. They are constantly licking her and passing it from bee to bee in a sort of messaging system. If the colony becomes too big, this dilutes the amount of substance being passed around. Likewise, if the queen gets too old, she will produce less. Once there is less substance in the colony, for whatever reason, it stimulates the workers to start producing the next queen.

They will make some bigger cells, and the queen will either lay eggs into these or, the workers will place fresh eggs in them themselves. When the new queen hatches after her diet of royal jelly, she will be paid little undue attention. It is only after she has mated with the drones that she suddenly gains their respect. Under natural conditions, it is now that one of two things can happen. The new queen might leave-taking a substantial number of the colony with her in what is called a swarm, or she can take command of the colony. If it is the latter, the workers will stop feeding the old queen, and she will either starve, or they will execute her. A hive cannot allow to queens to peacefully co-exist.

In the controlled environment of your hive, this process is managed by you, the beekeeper. If you time things right, you may be able to capture the bees before they swarm and place them in a new hive. Alternatively, you may want to replace the older queen with the new one, and then you will need to kill her so that half your colony does not leave.

Other than the queen and a handful of drones, all the bees in your colony are workers. They are smaller than the queen and drones, and though their individual roles are less obvious, they still play a crucial role collectively. The eggs hatch and are fed for the first nine days. They are fed a different diet to queens called brood food, which is just pollen and nectar brought in by the foraging workers. At nine days, the cell is capped, and the bee continues to develop for a further twelve days before hatching and starting to take up its role in the colony. It is interesting to note that the queen bee takes only sixteen days from eggs until hatching. This probably relates to the high-intensity diet and the fact that a queen is so important to a colony that they have evolved to reach adulthood more rapidly.

The worker does a number of different jobs over the course of its short lifespan of anywhere between five weeks and six months. The length of a worker's life is dictated by how much work he is required to do, and that, in turn, is dictated by the size of the colony and the pressures that it is under. For the first few days, the newbie is given a variety of newbie tasks such as cleaning empty cells or sitting on brood to keep it warm. If it is hot, it might be detailed to stand at the entrance to the hive and use its wings to act as a fan. Temperature control is one thing that is crucial to any hive.

It will be six days before the bee actually does any flying, and even then, it will start off with a series of short training flights rather than any big operational missions. Throughout their lives, the workers will keep progressing up the careers ladder and performing different roles. These include gathering pollen and nectar from incoming bees, transporting it to the brood chamber, making wax for capping, and throwing out dead bees or larvae. At eighteen days, they graduate and then move onto the important role of foraging for food and water. Quite how this career ladder works is still uncertain. We have some ideas, but the bees seem to also have a fairly relaxed attitude toward job titles and designation. A bee may forage for a few hours and then return to the hive and perform another task that would normally have been left to a newbie bee. If it feels like a bit of downtime, it might wander about the hive grooming the queen or passing nectar to other bees as they go about their work.

From long-term observation, we have a reasonable idea of the way the bee progress through the different work stages. The reason that it has been difficult to pin this process down more accurately is twofold. Firstly there is an element of flexibility in this progression which and it is impossible to put exact parameters on what is an inexact procedure. The second is that activities within the hive will vary according to the circumstance in which the colony finds itself. If a hive is thriving, then the steps above are more likely to be followed. If, on the other hand, the colony is struggling, then things will be changed by force of circumstance, and nurse bees may be forced to start foraging much earlier, for example.

Acquiring Bees

You've chosen your hive, assembled and positioned it, in what you hope is going to be the perfect position for happy, productive bees, but how do you go about getting the actual bees. This is the part you have really been waiting for. In fact, if you have skipped ahead to get to this section, please go back and read the previous chapters. There is stuff there you really need to know.

Obviously, you can't be a beekeeper without any bees, but acquiring a hive is not something you can do by just popping down to the local supermarket. In fact, there are ways to acquire bee colonies that aren't that different, but that is only one of several options. To start your colony, all you need is a queen and some workers. Believe it or not, you can purchase a colony (package) of bees that will be delivered to you through the post. What you get is a box with two sides of the screen mesh. Inside will be three pounds of bees, another small box called a queen cage, and a tin can feeder containing sugar water. There are between 10 000 and 11 000 bees in a package. This is an easy way to purchase bees, but there are some problems you need to be aware of. Firstly the bees may have come from some distance and are not familiar with the local conditions, so you are not sure how well they will adapt. Secondly, that journey through the postal system might not be too pleased for them, so before you even go through the process of introducing them to their new home, they are already stressed.

The second method is to approach a local beekeeper and ask if he will sell you a nucleus. Nucs, as they are referred to, are almost miniature hives with three or four frames and a small colony inside.

The third method is to capture a swarm. This is easier than you might think, but you have no idea when one will become available.

There is an old saying that says:

A swarm in May is worth a load of hay

A swarm in June is worth a silver spoon

A swarm in July isn't worth a fly

As you become known as a beekeeper, you will often be approached by wide-eyed people telling you that there is a swarm of bees that have just moved into their garden or roof, and they will want you to deal with them. For the time being, you are keen to get started, so options one or two are probably the best for you.

BEEKEEPING

Introducing your Bees to their New Home

To get the bees from the package box, place it near the new hive and remove three or four frames from that hive. Gently open the package and remove the food in and the queen cage. The queen cage is a small box about the size of a matchbox with two sides of the mesh. It is sealed with bee candy at one end. Check that the queen is alive and well. You can now hang the queen cage from one of the center frames in the gap you created by removing those three or four frames. The mesh must be facing the back or the front of the hive so that the workers can access the queen and are not blocked by the frames. Replace one frame on the other side of the queen cage so that it has a frame on either side of the queen.

BEEKEEPING

You now have two options. You can turn the package upside down and place it over the brood box containing the new queen. You then place another of you hive boxes over that and put the lid on. Over a period of about twelve hours, the worker will gradually make their way from the package box into the hive to be near the queen. They will slowly eat through the bee candy, and the queen will be released to carry out her duties. The following day you remove the lid, take out the package box and slide back the remaining frames into what is now an occupied hive.

A quicker way is to hang the queen cage as above, then turn the package over and firmly shake or tap it until the bees literally fall into the hive. In terms of time, there is a convenience element here, but if you don't need to bounce your bees about so violently, then why bother them. They have, after all, already had a fairly traumatic journey. About three days after you install your colony to their new hive, it is a good idea to do an inspection to see how they are settling in. The main thing you are doing on this quick inspection is checking that the workers have managed to eat through the bee candy and release the queen. You will easily be able to see because if she is still in the cage, it will be covered in workers. If not, there will probably only be six to a dozen bees on the cage. Pick up the cage slowly and look inside. If it is empty, then all is well. Shake off any workers and take the cage out of the hive. If she is still inside, then you can carefully pierce the candy and make sure she is able to get out before either shaking her gently into the hive.

Transferring your new bees from the nuc to the hive is even easier. Remember that it is effectively a miniature hive. Open the empty hive and remove one more frame than the number of frames in the nuc. Next, you open the nuc and gently remove the first frame bees and all. You will probably need a beekeeping tool to lever the frame from the nuc because the bee will have stuck it down with propolis. Propolis is a dark sticky glue-like substance that the bees make and then use to seal all the little cracks and crannies to keep their homes' weather tight. Because the nuc frames are the same size as the frames of your hive, you now simply slot the first one into the first empty position in the hive. You continue the process until all the frames from the nuc are in the hive.

There are a couple of important factors to bear in mind here. Place the nuc frames in the same order in the hive as they were in the nuc. This way, the bees are familiar with the layout, and this will cause them the least possible disruption. Also, on one of the frames, you will find hanging the queen cage. This should now be empty as the workers will have released her, and she will hopefully be laying eggs into one of the frames already. You can remove the empty queen cage.

Once you have installed the nuc frames to your hive, you will still have many bees in the nuc. Turn it over and shake those bees into the hive. You don't want to simply sit it on top of the hive and cover it with the next box as with the package because if the queen is still in the nuc, the bees won't move down into the hive. Don't worry if not all of the bees fall out of the nuc. There will always be some that cling to the edges. Put the lid on the hive and then place the open nuc right in front of the entrance. The bees that have been left behind will smell the pheromones from the queen and gradually make their way into the hive of their own accord.

Finally, we come to introducing a swarm that you may have been lucky enough to capture. Very often, the swarm will have clustered on a branch, and you will have been able to cut it off and place it, bees and all, into a box or some other container. A pillowcase is quite a commonly used method. Take an ordinary bed sheet and lay it from the entrance to the hive to the ground in front to form a sort of ramp or welcome carpet, if you prefer. You then simply place the swarm on the sheet, and the bees, now anxious to find a home, will make their own way into the hive. An interesting thing to look out for as this happens is bees leaning forward and waving their abdomens in the air. What they are doing is secreting a hormone from the Nasonov gland, which acts as a guide to the rest of the swarm as to where they should be heading where the queen is.

Bees will swarm from April right through to August. Bees swarm when they feel that they have reached a stage where the colony is so healthy that it can split. There is a recognition that with such a large colony, resources will become scared, and so a percentage of the colony moves on in search of a new home. For the novice, the sight of a fast-flying cloud of bees can be quite intimidating. The bees are not normally aggressive at this stage, however. Prior to swarming, they will have gorged themselves, and they will not have a hive that they are protecting. The problems normally occur when they come to rest, which can be in the most awkward of places. They may stay in one place for as long as two or three days after that, and if it is somewhere that is close to pedestrian traffic, the risk of people getting stung does increase, especially as the bees use up their food reserves and start to get hungrier. The place that the bees gather is just a staging post while scouts go out and look for a suitable place that they can turn into a permanent home. Beekeepers often leave an empty hive in the hope that this will attract scouts, and they will then guide in a homeless swarm. Bees that swarm often tend to repeat swarm because it is part of the queen's character. To overcome this, when you capture a

BEEKEEPING

swarm, you should replace the queen as soon as possible.

Whichever method you opted for, well done, you are now a beekeeper with all the pleasures and responsibilities that accompany the position.

CHAPTER 6

HIVE MAINTENANCE AND TREATMENT

From now on, we move onto the maintenance side of beekeeping. The learning curve that you have just been through has probably been quite steep if none of this was familiar to you. Things will get easier now, and in many ways, you have crossed a threshold. That early information acquisition is quite large, but now things will start to take a more leisurely and soothing tempo. You are working with nature, which brings with it its own rhythms and pace.

Under normal circumstances, you want to check on what is happening in your hive or hives, about once a week. These checks are an important part of bee husbandry. It lets you know what is going on in the hive and enables you to respond if there is any sort of a problem quickly. Those inspections will also make you more familiar with open the hives and let both you and your bees get used to one another. This is important because it gives you confidence. Many beekeepers believe that the bees can sense when you are stressed and that this rubs off onto them and affects the way in which they behave. Whether or not you subscribe to this belief, the more you interact with your bees, the better beekeeper you will be.

It has always been taken for granted that the months during which you performed weekly inspections ran from April to the end of September. Changing weather patterns mean that we can no longer make that assumption. You will need to observe local conditions and respond accordingly. If it is warm autumn, the bees will continue searching for pollen and nectar for as long as possible, and that may mean you need to make some changes later in the season than would have been normal just a decade or so ago. The best time of day to do your inspection is from around eleven in the morning until four in the afternoon during periods of fine weather. The reason for this is that most of the workers will be out foraging at that time so you will be disturbed as few bees as possible.

Initially, when doing inspections, you will probably want to wear a bee suit. Over time you will become more confident and more accustomed to getting stung, and you may then be able to do away with some or all of this protection. You will also be better at reading the hive and assessing whether they are agitated or not. Regardless of what regalia you opt for, be aware that strong perfumes can annoy bees, so you might want to skip your regular ablution routine on inspection days.

Before approaching the hive, to get your smoker going properly and have your hive tool handy. If you are going to be feeding, then have a feeder prepared. Now move slowly up to the hive and give a couple of puffs at the entrance before moving to the rear of the hive and removing the lid and placing it upside down on the ground. Carefully lever opens the top cover, which will have been glued in place with propolis. Blow in two or three puffs of smoke, then replace the lid for a second or two so that the smell of the smoke permeates the hive slightly. Bees are not calmed by smoke, which a popular misconception. Instead, they assume that a fire is approaching and the gorge themselves with honey so that if they are forced to flee, they have some emergency supplies on hand. While they are gorging, they are too busy to pay attention to what the beekeeper is doing. After one or two minutes, open the top cover again and place it upside down in the lid.

You are going to start your inspection from the bottom box of the hive and work your way upward. Standing at the rear of the hive, take off the top super and sit it on the cover and then repeat with any other supers and then the top box placing one on top of the other in reverse order to how the hive originally was.

When you are down to the bottom box, then carefully lift out one of the frames. It will be well stuck down, so use the corner of your hive tool to lever it lose. Take the frame and examine it carefully. You are looking for capped cells, cells with larvae in, and most importantly, eggs. If there are fresh eggs, then you know that the queen is around. A healthy queen will only lay eggs in the center and at the bottom of a cell. It would be nice to see the queen, but you don't want to spend too long looking for her, and as long as you see fresh eggs, you know she should be there somewhere. You are also going to be looking for parasites such as wasps, moths, or beetles, but we will go into this later when we deal with the subject of hive pests.

Lift out an edge frame. You always work from one side to the other to maintain the original order of the frames. Once you have inspected it, then hang it from frame hooks on the side of the hive. This leaves you a touch more room when taking out the other frames. You will continue with the second frame, which, once inspected, you will place directly back into the box. You continue like this until all the frames have been inspected. This ensures that each frame goes back into its original position. When you have checked the last frame in the bottom box, take the first frame off the hooks and replace it in the position it was when you removed it.

It is important not to rush this process. Move slowly and be gentle, and everything will be fine. Take a minute or two to check that all is well and then be careful when replacing each frame. They should be crawling with bees, and you don't want to harm any of your stock. You definitely don't want to crush your queen accidentally. In a full hive, you may want to have a bee brush handy so that you can brush away some of the bees if they have gathered into a clump while you are doing the inspection. This just helps prevent crushing and makes sliding the frame back into place easier. Generally, it is not necessary. When you have checked all the frames in the bottom box or deep, replace the one that was above it. When placing a box or super on a lower box, use a sliding action rather than trying to drop it directly into what will be its final resting place. This is because there will be bees sitting on the lip by now, and you want to push them out of the way rather than crush them between the two boxes.

At the end of the inspection, all frames, supers, and boxes should be in exactly the same position as they were when you first opened the hive. You can now replace the top cover, fit a feeder if required, and close the lid.

There will be a difference in the bottom boxes to the top supers. The brood is in the bottom, so that is where the eggs and larvae should be. There is normally a guard to prevent the queen from getting into the supers so that she does not lay eggs where you are going to gather honey from. The frames in the supers will be lighter in color as they only have honey in them. It is always a good idea to hold them up to the light to check for eggs simply to ensure that the queen has not been able to sneak up above the brood boxes. In the brood boxes, you will be looking for the queen. It is always nice to see her, but don't let it make you waste time. You want to keep the disturbance of your stock to the minimum. As long as there are fresh eggs, you have a pretty good idea; you have a healthy queen. Bees only stay in their egg form for two to three days before becoming larvae, so that gives a fairly reliable timetable as to when the queen was last there. She also tends to remain in the same vicinity when she is lying, so if you do need to find her, that is where she is most likely to be.

Another thing that you will be keeping an eye out for our new queen cells. These cells are much bigger than ordinary bee cells and cause an almost blister-like a bulge on the comb. Worker bees produce these so that a new queen can be reared. Sometimes they just seem to make them a sort of reserve should they need to produce a new queen quickly. On other occasions, they do it because they sense that the existing queen is getting too old, and that means they are preparing to replace her. When you come across a queen cell, check it carefully. Providing it has no egg inside and is not capped, then you can ignore it. If it looks like a new queen is on the way, you will probably need to take action, and we will look at the procedure you follow a little later. While you are in the bottom boxes, you will be able to gauge what the food reserves are like and decide if there is supplementary feeding needed. Normally queens lay in a pattern, and this will show if she is performing well. You will also see if the brood looks healthy and if there is crowding starting to take place. If there is, you may need to add a super or even think about splitting the hive, which we shall get to later.

Inspections are important but don't get too carried away. Every time you open the hive, you are creating a disturbance that slows down production and adds stress to your stock. Some keepers estimate that opening the hive is the equivalent of one day's production lost. Checking every seven to ten days during the warmer months will be perfectly adequate, providing no problems are detected that demand further action.

It is rare to inspect hives during the winter months. In the colder months, the bees huddle up to keep warm, and opening the hive instantly causes it to lose temperature. Temperature control is one of the primary requirements of a healthy hive.

Record Keeping

Record keeping might not seem important, but it really helps you become a better beekeeper even if you only have one hive. In future years you will be able to look back and see how the hive performed, but also how you responded and whether there are things you could have done better.

There are several ways you can keep your records. Today some keepers use video, and others have a small portable voice recorder, but most keepers rely on simple old-fashioned notebooks. You might think you will remember everything, but it is highly unlikely, so try to take notes soon after closing up the hive. If you end up with more colonies, the need for records becomes even more important.

First of all, have some way of designating one hive from another. Next, note down the overall condition of the bees. Were there plenty of them, how did they look, were they foraging well. Did you see the queen? Were there plenty of fresh eggs? Was there a healthy-looking brood pattern without an overabundance of bullet brood? Bullet brood is larger, darker brood cells that indicate they contain a drone. If there are lots of them, it shows that the queen has been lost, and an unfertilized worker has taken over the laying. How much food was in the hive, and does it need to be supplemented. Does it look like the hive is preparing to swarm?

All of these are details that will help you look back and evaluate both your own beekeeping efforts and how the hive performed. Each inspection will give different results at different times of the year. What you see in the hive in late autumn will differ very much from what you see in the early summer.

Seasonal Feeding

Making sure that your stock has enough food is critical. Generally, you supplement food prior to the winter months when there are no flowers, and the bees are in lockdown for the winter. There may, however, be periods when you need to supplement their food supplies outside of the winter months, and one of the things your inspections will tell you is how strong their reserves are.

Bees should be given food from late autumn. They will transport this food down into their hive, where they will store it and be able to access it without leaving the hive and letting in the cold. There are numerous bee-feeding options, but here are some of the easiest and most common.

A feeder frame can be hung in the hive. It is the same size as a normal frame and can hold one gallon of sugar water, which you can mix yourself at a ratio of 2 to 1 to create thick syrup. This is a really easy solution to feeding, but it holds only one gallon, and the average hive will need four gallons to get through the whole of winter. This leaves you in the unenviable position of having to open the hive several times to fully stock the colony at a time when the days are cooling fast.

An easier method is to top feed with a feeder bucket. The bucket is simply a bucket with a lid that has a hole in it. Open the hive and place a feeder board on top. This is simply a board like the top board that had a hole in. The bucket is then inverted, and the syrup can escape. It forms a vacuum, so you don't have to worry about the whole container dropping into the hive and drowning the bees. As soon as they realize that it is there, the bees will come up from the hive and start gathering it in a process that will take around four days.

Another option is the top feeder. This is almost identical to a super, but inside is a solid floor with two panels that effectively turns it into two separate containers separated by a slot. Fill the two containers with syrup, and the bees will come up and start to help themselves. When filling the feeder, dribble a little syrup into the hive, and this will speed the time it takes for the bees to realize that there are easy pickings to be had right above their heads. Once the feeder is full, lay a few handfuls of straw on top of the syrup so that the bees can get to the food without drowning. You then close the lid, and four days later, the feeder will be empty, and you can open up and simply remove it.

The final option if you have several hives is to fill a standard forty-four-gallon drum with enough syrup to provide four gallons per hive in the vicinity. Prop the lid open slightly so that the bees can get inside and dribble a little syrup around to alert them to its presence. It won't take them long to find the food and start carrying back to base. This is a really straightforward method if you have several hives near to one another. If there is a risk from predators that might try to steal it, you will, of course, need to take the appropriate measures to prevent that happening.

Like so much in beekeeping, what sounds quite complicated at first, turns out to be really simple. Remember that the bees do most of the work.

Below is a seasonal checklist for you to follow now that your hive is installed.

Spring

At the moment, your hive is new, so they will focus their efforts on filling the hive and building up their supplies. Next year, however, it would be a good idea to have an empty hive nearby so that if they do swarm unexpectedly, there is another home that you will entice them not flee. By then, they will have started to gather food from the abundant flowers in the area, and there is little chance of them starving, so it would be a good time to harvest the honey. We will look at how to do that later in the book.

Summer

If you are still supplementing their diets with bee candy or syrup, you should be able to stop now. Make sure that stronger hives are not robbing yours. Continue your regular checks keeping a close eye for verroa mites. If you do not have a permanent source of water nearby, then make sure you place somewhere they can find it. You will discover the bees are thirsty at this time, especially if it is hot. Later in the summer, you will need to assess if there is enough honey to harvest.

Autumn

If you have more than one hive and one is looking particularly weak, then you may want to combine the two. Check that there is food available if flowering plants go over, and there is nothing to harvest. Protect hives from the wind, but don't forget that there must still be ventilation. The entrance can be fitted with a mouse guard and narrowed to allow for reduced use by the bees. This will help it stay warmer in the hive. Don't harvest any honey unless you are sure that they will have adequate supplies to get them through the lean winter months. It is also a good idea to weight down the roof so that it cannot be blown off.

Winter

In the winter, you are going to open the hive as little as possible. You can take the lid off on warmer days just to ensure that they have enough food. Your main jobs at this time of year consist of checking that hives are not threatened by winds.

We have looked at the ongoing jobs that need to be done on an ongoing basis. Now we will look at other tasks that are essential to beekeeping, but which occur less regularly. Some of these we have touched on earlier in the book, but now we will look at them in greater depth.

Swarm Prevention

As a beekeeper, you want your colonies to become successful at food and brood production. The problem with this is that when this success starts to be achieved, the bees' natural instinct is to divide and swarm. This can result in the loss of a large portion of your stock. Up to sixty percent of your bees might just up sticks and fly off accompanied by the old queen. Your job is to circumvent the colony's natural instinct.

Some colonies are more likely to swarm than others. We know that much of the bees' behavior is influenced by the queen, and if you have a captured swarm, then it is highly likely they will swarm again. They won't normally do this straight away, however, so there are advantages in taking a swarm if you can and if they have installed themselves in a place that would require a trapeze artist to get to. Swarms are keen to get their new homes in order, and so they are really zealous when it comes to building new comb and producing a brood. They can be left for a while so that you can profit from their diligence. After that, you will have to replace the queen with one less inclined to swarm. You can also combine the colony with any weaker hives that you may have. When a swarm has settled into one of your hives, pay particular attention when you do your inspections. You must ensure that they are not introducing disease or verroa mites.

If there is plenty of space in the hive, then there is less likelihood of swarming, and you can control this with the addition of new supers before crowding becomes an issue. A young queen is unlikely to lead a swarm if she is in her first year of egg-laying.

Some stock is better at staying put than others, and once you get to know your bees, you will start to recognize which ones pose more of a flight risk. You will then want to breed from that colony and expand that stable characteristic. One way of knowing which colonies are keen to swarm is by checking for queen cells on your weekly inspections. If there are very few, then they are probably not looking to divide, but if the number starts to go up, then beware. Normally after August, the bees will not swarm. It would be very late in the year for them to establish enough food to get them through the winter.

Artificial Swarming

This is a procedure that can be extremely useful for preventing your bees from swarming while at the same time creating another colony. Here we need to take advantage of the fact that bees are extremely accurate at returning to their hive. If you move the hive only a few yards, they will need to search for it again. If you move the hive and replace it with an empty one, they will think the new hive is their original home. During the warmer part of the day, when most of the workers are foraging, move the old hive about two yards and place the new one in position. After that, open up the old hive and locate the frame that the queen is on. You then place this into the new hive along with a full box of frames. The retuning workers will find the queen, several empty frames, and some brood on the frame from the old hive. This is more or less where they would have found themselves had they swarmed. They will immediately set about drawing out the new frames and basically expanding into their new home.

The old hive is now occupied by a number of bees, many of whom will have never made their maiden flights as well as capped cells that will soon hatch. These bees will leave the hive and fly out and then return to that position, none the wiser to the fact that they have been transferred. The problem you have is that there is no queen in that original hive. You now perform an inspection looking for queen cells and in particular, queen cells that may already contain an egg. All of these you will destroy by crushing with your fingers or your hive tool. The workers will quickly sense there is no queen and will immediately set about making new queen cells. Into these cells, they will start placing eggs that are already in the hive and start feeding royal jelly.

We know that it takes fourteen days for a queen to hatch. What we don't want is to find ourselves in a position of having several queens hatch at the same time. To avoid this, after seven days, you reopen the hive and destroy all but the most healthy-looking queen cell. If all goes well, seven days after that, when you do your inspection, you should find that you have a fully functioning colony under the leadership of a brand new queen. You have just doubled your hives and hopefully prevented swarming in the process.

Uniting Hives

There are a few reasons that you may need to unite two colonies to form one functional hive. The most common reason would be that one hive has become weak for some reason, and it seems unlikely that it will survive the winter. By combing tow hives, you give both colonies a better chance of survival. Another reason might be that you have captured a swarm, and you have decided to kill the queen to prevent future warming. A third reason might be that you have lost a queen for some reason.

Uniting two hives is a straightforward procedure, but there are some factors that you must bear in mind. Bees will smell if bees from another colony enter their hive and will attack them as they assume they have come to steal their honey. There is a simple way to overcome this. First place both hives beside one another. As long as the bees are not moved more than a yard, they will easily find their way back to the hive you want them to live in. Next, open both hives. Over the stronger hive brood box lay some sheets of newspaper, then place the queen excluder over the paper and score the paper lightly in one or two places. Now search the frames of the weaker hive and find the queen who you will kill by squashing her between your fingers. Only once you are sure that you have killed the queen should you place the second brood box on top of that which is covered in paper. Place the supers on top of that and close the hive.

The lid and bottom tray of the hive you have moved will undoubtedly still have bees crawling all over it. Just knock these guys out onto the ground, and they will eventually make their way back into the hive. The bees will take a couple of days to chew their way through the newspaper barrier you have imposed on them. In the process, their odors will have mingled so that by the time they can get through, they will be used to one another. You will see small pieces of paper lying outside the hive, and this will let you know that they are doing their job and chewing through the paper.

BEEKEEPING

In other types of hives such as top bar hives, this method cannot be used as the hive cannot be broken down into component parts. All is not lost as there is another way of getting bees from separate colonies to get along with each other. Spray the bees from both colonies with sugar syrup and then shake the bees from the smaller colony into the stronger one. The bees will be forced to lick one another to get clean, and by the time they have achieved this, their odors have mixed, and they accept one another.

There is yet another angle to combing bee colonies, and that is when you have three colonies that are too weak to survive over winter. By shaking all the bees from the three hives into one hiver, the combination of different colony smells so confusing that it overwhelms everyone, and they just seem to get on.

It is most important that whatever method you decide to use, you only end up with one queen in the final hive. It is difficult to kill a queen at first as it goes against all beekeeping instincts and seems almost counter-intuitive. What you must remind yourself is that if the queen does not die, the whole colony almost definitely will.

Pest and Disease Control

One of the biggest problems being faced by beekeepers today is the management of pests and diseases. There are many reasons that these problems are now so pressing. One of them is that we have transported bees from different parts of the world, and in the process, we have introduced these pests or diseases to areas where they were not experienced previously. Add to that the fact that many commercial operations transport vast numbers of bees over long distances to be used as pollinators, and you can see why the problems are being exacerbated. Transporting bees stressed them, and stressed bees are more susceptible to health issues.

Pest and disease management are some of the primary reasons why you need to do your inspections diligently. Before you can identify the problems, you need to be able to recognize what healthy bees and brood look like. After that, you will quickly be able to spot any anomalies. Brood should be a uniform light brown color and joined together in what is called pattern rather than dotted about here and there. There should also be a few empty cells within the pattern itself.

Verroa Mite

Verroa is a small red or brown mite that has caused dramatic problems in the beekeeping world and mass casualties in wild colonies. The mites attach themselves to bees and suck their bodily fluids. This in itself is not the problem as the bees seem to carry on regardless. It is in the cells where they breed that most problems occur. This is the most widespread pest for beekeepers. It is thought to have originated in Southeast Asia and has spread to most countries other than Australia who have introduced strict control measures to try to prevent its arrival there.

The mite is visible to the naked eye, and once you have an infestation, you will probably see bees working in the hive with what looks like a brown dot attached to their thorax. Although the problem may seem insignificant, it will augment dramatically when the colony size decreases over the winter months or if there is a sudden drop in bee numbers for some other reason. At the moment, we are still not sure to what degree this mite has played a role in colony collapse disorder, but wild colonies seem to swarm in an effort to escape the mites, and then they simply disappear. To monitor the size of the infestation, a special sticky paper is fitted to the bottom board. This holds the dead mites that drop down, and after a seven day period, the beekeeper can work out from there an approximate rate of infestation.

There are many different methods of reducing mite infection, but so far, no total cure. The objective at this stage is to keep their numbers as low as possible rather than attempting total eradication. The first method is through the use of what are called hard chemicals. They are mainly pyrethroid based, but they come with some problems. Firstly they need to be used during periods when honey is not being gathered to avoid cross-contamination. Secondly, the mites have started to adapt to them, and they are becoming less effective, and thirdly, many beekeepers are averse to the use of any toxic chemicals in their hives.

Soft chemicals based on thymol are another option. These are the bee equivalent of homeopathic medicine. The final option is that of mechanical control. Mites are attracted to drone brood to lay their eggs, and so if bullet brood is destroyed just before they hatch, the mite numbers will be reduced. The mites also like to lay into uncapped cells, so it is possible to restrict the queen in a limited range for a short period. The frame she lays in during that time is then sacrificed, thus helping to reduce the infestation further.

The verroa problem is far from being solved, and at the moment, the most realistic option is to try to manage it as well as possible. This can incorporate one of the solutions above or an amalgamation of them. Yet again, liaising with nearby beekeepers will encourage a transfer of information that can prove invaluable. This problem has been around for some time, and you should not let it put you off keeping bees.

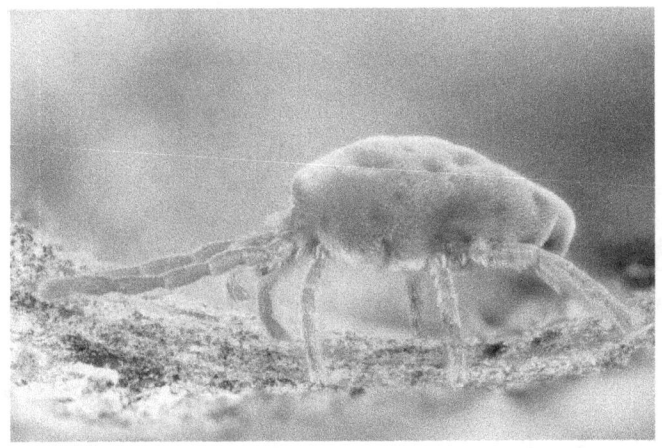

Tracheal Mite

These tiny mites are invisible to the naked eye. They inhabit the breathing tubes of the bee, where they suck bodily fluids until ready to breed. At this stage, they exit the bee and crawl up onto one of the fine hairs of its upper body. There they wait to climb onto a younger bee, enter the breathing tubes again, and lay their eggs with the intention of repeating the cycle.

The control consists of placing vegetable shortening patties mixed with sugar onto the top bars of the hive. When the bees come to lick the patties, they come into contact with the shortening, which prevents the mites from detecting the young bees onto which they need to climb to complete their life cycle.

Small Hive Beetle

This is a small round black to dark brown beetle about half the size of a bee. They were first detected in Florida in 1998, is thought to have originated in Africa. By 2006 they had spread too many parts of the US. They move into a hive where their larvae can turn comb into dark slimy goo. The bees are unable to tote them through there tough carapaces, and though a strong hive can tolerate them, it is the smaller weaker hives that are most at risk. If they become too established, then the risk is that the bees will swarm and abandon the hive altogether in an effort to find a clean pest-free environment.

The larvae normally establish themselves in the lower corner of frames, so you should look out for them there. They are also attracted to honey, so your honey house may be at risk if you extract honey and don't deal with it soon afterward.

The only chemical that the beekeeper can use is the same as those for verroa. The pupa stage of their lifecycle is performed in soft sandy soil, so if you have any of that near the hive, you may be able to treat it but beware of what chemicals you use. There are traps sold commercially that allow the beetles to enter but are too small for bees. The traps are baited, and the beetle drowns, normally in some sort of vegetable oil, after entering. These traps work to a limited extent but will not succeed against serious infestations. The main defense is to keep colonies strong and take action early if beetles are detected.

Greater Wax Moth

There are two types of wax moth, and both are a problem to the beekeeper because they live on wax, stored pollen, and the remains of bee larvae. The male moths attract female mates by emitting ultrasonic sounds. This causes the female to approach, and as she gets closer, the prospective mate will release pheromones to bring her closer enough to mate with. The female enters hives and lays her eggs within tiny cracks and crevices. Once the eggs hatch, the larvae start to eat their way through the comb. At the same time, they produce a fine silk web that can trap emerging bees. In stronger hives, they are not generally too great a problem, but they can devastate weaker hives. It is crucial to practice strict hygiene control in an effort to eliminate these pests. None of the stages of the wax moth lifecycle can withstand extreme cold, and what some beekeepers do is freeze empty frames for a couple of hours to kill any eggs that may have been hidden on them.

American Foulbrood

If verroa is by far the widest spread and most common pest, then foulbrood is its disease equivalent. American foulbrood (AFB) is a highly infectious disease that affects capped brood. The caps of the cells will be dark and wrinkled, and the brood pattern will be erratic. In more advanced cases, there will be an unpleasant odor coming from the hive when you open it. One test is called roping out. This involves sticking something like a match stick into one of the cells. As you slowly withdraw it, a gray sticky mucus will cling to the match stick, and this indicates a strong likelihood of the disease presence. In some of the cells, the larvae will have already become desiccated, and if you look down into it, you will notice a dark shiny object at the bottom called scale. There may be live larvae interspersed among the affected larvae.

If you believe you have an infection, you should ask a more experienced keeper to look at your bees for you. AFB is a notifiable disease, and once you are sure you have it, you will need to notify your regional bee inspector or another governing body. They will oversee the destruction of the hive, which will probably involve burning it. You will also need to pay extra attention to cleaning all your equipment as the spores of the disease are highly infectious.

European Foulbrood

EFB is an infection of the uncapped cells. As with AFB, there will be a lack of healthy patterns to the brood. Inside the cells will be distorted larvae. Instead of the usual white c shape, they will be twisted and deformed. The disease creates a UT infection that ultimately kills the pre-emergent insects. This disease is less dangerous than AFB but can still be serious. Again you should notify the relevant authorities that will inspect. Depending on the severity of the case, they may offer options that could include ant biotic treatments. You will need to discuss your options with them, but most experienced keepers would still prefer to destroy the hive than risk having the disease spread.

These are just some of the most serious disease and parasitic threats to you bee stock. There are plenty of others, but the important thing is to be aware without being put off. You are dealing with nature here, and anything natural comes with both wonders and threats. It would be no different if you were growing vegetables or raising chickens. All of these problems can be managed if you are aware of them and keep doing your inspections. The most important factor to take away from this is that building up strong, healthy colonies almost always enables the bees to overcome the problem. More than anything, this should be your main objective as a beekeeper. As you become more experienced, you will gain confidence, and this will be just something that you both manage and take in our stride.

CHAPTER 7

HARVESTING THE HONEY

In this chapter, we will take a look at what for most beekeepers is the most enjoyable part of beekeeping, bringing in the honey. We will also take a brief look at some of the other products that can come from your hives. Of course, honey is by far the best-known bee produced item, so let's start by taking a closer look at the different types of honey.

Honey

You will probably be surprised to learn that there are in excess of three hundred different types of honey. As a rough rule of thumb, darker honey tends to be stronger while the lighter honey is sweeter. Honey is made up of roughly eighty percent fructose, and glucose, and the remainder consists of nectar and other natural products. When buying from an artisan, you will normally get raw honey, which has been filtered but not processed beyond that. In your local supermarket, you will mainly find refined honey, which has been through a heating process prior to bottling.

BEEKEEPING

This gives greater consistency, but at the same time, you lose many of the nutrients and vitamins of the raw product.

There is another reason that you may want to consider buying your honey from an artisanal producer. In the US, honey consumption had doubled. This is in part due to the fact that people are becoming less inclined to eat sugar due to a growing awareness of the health implications. Honey has become the go-to substitute of choice. The consumption may have double, but production in the US is down thirty-five percent. That gap may partially have been filled by imports, but there is another more sinister factor. Honey has become the third most adulterated natural product. In fact, honey fraud is big business. It is very easy to knock out a chunk of real product and replace it with sugar. Inspectors are getting better at looking for this type of fraud, and hopefully, it is not going to affect you for long because you are soon going to have an abundant supply of your own home-produced raw honey that you know has not been tampered with.

You will often hear honey types being referred to by the type of flower from which they were made. Acacia honey, for example, is made from the flowers of the locust tree and is quite thin and pale in color. Honey from ivy tends to be thick and strong. The reason beekeepers are able to determine the type of honey they are selling is that we know that bees will continue to return to the same source flower for as long as it is available. This makes sense because if the bee has discovered a good source of food, then why should it waste energy searching for another one. In some cases, the supply of flowers might be more mixed. This is often the case with spring-flowering wildflowers or urban honey, where the source is more varied.

The type of flower from which honey is produced is not the only factor that will influence the flavor or even the texture. Honey flavor can be influenced by whether it was raining or not, temperature, and even may differ from hive to hive. It is that slight element of the unknown that makes the honey harvest so interesting.

Besides temperature, there can also be differences in texture. Some honey is thin and runny, while others are thick. This often results from the flower being harvested by the bees. Honey that crystallizes more readily do so because they are higher in glucose than in fructose. They also lose flavor more readily. Ivy honey can become so thick that the bees themselves are unable to access it as a food, and you might night to supplement them even though they are bringing in vast amounts of stock.

Honey is mainly harvested between late July and mid-September. In areas where there are plenty of flowers, and the weather is kind, there may be a harvest in July and the second one in September. The bees will cap off the cells, and this indicates that the honey has a watery consistency that is low enough. The capping helps to keep the honey warm.

When you open the hive, it is likely that the top super is not full enough to warrant taking the honey. Place this in the lid, and the next supers down should be full enough to warrant removing the frames. There are three different options here. The most popular is to take out a frame and brush of the bees using a soft bee brush. Once the frame is clear of bees, you place it in an empty super and put a lid back on, or the bees will start to reclaim their honey. You work your way through the super in this manner until it is empty and then move down to the next super if you have one.

After you have removed the honey down to the brood box or deep, you put what was the top super back on to it, so the bees have somewhere to carry on working if you are expecting good production to continue then you can get another empty super on top of that.

Another method that is used by beekeepers with many hives is to stand the super that is to be emptied on its edge and blow all of the bees out with a leaf blower. By blowing between the frames, it is quickly cleared of bees and then can be covered before you move onto the next super. This method is quick and efficient and means you don't need to force the frames out while the bees are still in the box but can do that in more comfort back in the honey house. In terms of efficiency, this method works very well, but it is not for you if you took up this hobby to enjoy a quiet time interacting with nature.

The final method of clearing the frames is to fit an escape board. This board is like a top board in that it fits precisely over the deep or super. It differs in that it has a hole in it through which the bees can get out but not back in. Once all of the supers have been removed, place an empty one, complete with frames onto of the deep, cover it with the escape board, and replace the filled supers and finally the lid. In the evening, the workers make their way down into the brood box to keep the brood warm with their body heat. Over the course of about two days, the upper supers should be empty of bees, and you can remove the full supers and the escape board and take them to the honey house.

Whichever of these three methods you opt for, seal any gaps between the supers with painter's masking tape so that no robber bees can get into the hive. The smell of the exposed honey will have told them there are easy pickings to be had. Under normal circumstances, these holes are filled with propolis during the course of the workers day to day activities.

BEEKEEPING

By cracking apart the supers, you have left them exposed to bees from other hives. They will quickly set about resealing, and the tape will become redundant and can be removed.

Now you have several frames that are filled with honey and capped with wax. You need to get these to your honey house, or a room dedicated to this process. Some people use their kitchen, but you need to have all entrances sealed with mosquito screens, or bees will get in. They are attracted to the small of all that honey and are keen to get it back. The other disadvantage of not having some sort of dedicated space is that things do get sticky no matter how careful you are. If you think bees can get angry, just wait till your wife gets home and finds her once-pristine kitchen covered in sticky goo. No matter how many jars of fresh homemade honey you show her, you are going to get you out of trouble.

You are going to need a honey extractor. These come in varying styles, but most of them follow the same principle. They consist of two stainless steel tanks that sit on top of one another. The top one contains some baskets or racks for carrying the frames and some method of making them spin. This is usually as simple as a handle that you turn. More advanced extractors have a motor to do the turning, but unless you own dozens of hives, this is really not a requisite. The honey is then thrown against the sides of this tank

BEEKEEPING

by centrifugal force. From there, it runs down into the lower tank, which is little more than a large receptacle with a tap on it from which you pour the honey when you are ready.

Before loading the frames into the extractor, you will need to decap them. This is another very simple operation. All you need is some sort of deep tray to catch them in and a sharp serrated knife such as a large bread knife. Stand a frame on its end in the tray and then slowly slide down the frame using a gentle sawing motion. Use the edges of the frame as a guide to slide along so that you cut the wax off evenly. It should just slowly peel away and drop into the tray. In some places, the wax capping may be a little lower than the edge of the frame, but you can just scrape those caps off with a fork or the tip of a knife.

After removing the caps on both sides of the frame, you slot it into the extractor and repeat the process with the next one. How many frames you decap is dictated by how many the extractor will hold. Once all the racks in the extractor are full, just slowly but steadily spin the handle. You can check every now and then to see there is no honey left in the frames and just give another spin if there is. Between the two stainless steel tanks, there is a fine filter that prevents any small pieces of wax or other solids from falling into the lower tank. You repeat this process until you have spun all of the harvested frames.

It will take about twenty-four hours for the honey to run down the sides of the top tank into the lower tank. At this stage, you pour into jars directly or into honey buckets. Honey buckets are just bulk storage buckets with a tap on so that you can decant to jars at a later stage. You have just harvested your first honey. It is a good idea to let the honey sit for a day so that the bubbles can rise and escape. They will not harm the product, but they form a white film on top of the jar, which some people will not like. Don't fill jars right to the top but leave a gap of about three-quarters of an inch before sealing. Try to be consistent so that all of the jars are filled to the same level. The jars themselves must be sterile, and even if you are using brand new jars, you are well-advised to sterilize them first. Other than that, the main risk to your product comes from having too high a water content. Anything above eighteen and a half percent and you risk the honey fermenting and going off. Excess water comes from cells that are uncapped and still more nectar than honey. The bees are adept at only capping once the water content has gone down enough because they are instinctively aware of the fermentation risks. You will only have a problem if you harvest too much-uncapped honey. You can check water content using a hydrometer if you are uncertain, but with experience, you will be able to tell just from looking at the honey itself.

You are now left with a large bowl or two of honey-covered wax. Stand this in a sieve or colander and let it drain into another bowl. Again this takes around twenty-four hours. What you will be left with is a large bowl of honey and several pounds of wax. The honey can be poured into the extractor to filter through to the remainder of your harvest. The wax can be used for candles or incorporated into other products such as cosmetics or furniture polish. There are a number of sideline products such as these that can be added to your product range. Before using it, melt it on the stove at very low heat. Pour the liquid honey into a cardboard milk container or old ice cream container and let it harden. After that, it will come out as a block. At the bottom will be a dark mass, which consists of pollen and propolis. This you should cut off and what is left is virtually pure wax. Beeswax candles are favored by some people because of their natural smell and the fact that they do not produce any smoke.

Another option for wax is to swap it with the manufacturers of foundation sheets. Some beekeepers put sheets of this wax into empty frames so that the bees do not need to waste time and energy making comb. All they do is draw out the wax sheet and turn it into cells. Producers will take your raw wax and then give you a reduced price against the foundation sheets. If you have extracted your honey without damaging the wax cells, you can replace these into the hive, and the bees will simply refill them so that the foundation sheets are not required.

Many beekeepers share an extractor. They are one of the most expensive pieces of equipment, and they are only used once a year, so it makes sense to use them co-operatively. Most beekeeping clubs have one or two that its members can borrow. If for some reason, you are unable to get access to an extractor, there is another means of accessing the honey. You can cut the caps off as above and then just allow the honey to drain into a large tub or bucket. It is a much slower method and still needs to be filtered, but it does work.

The last option is simply to cut the comb up, place it in containers, and sell it in the comb. You will be surprised how many people prefer honey this way. It tastes the same, but it looks more authentic, and customers can be sure it has not been adulterated or processed.

Other Products from your Hive

Wax and honey are the two things that leap to mind when we think about bees, but they are far from the only products that a beekeeper can utilize or sell.

Propolis

We have mentioned this product a few times during the course of this book. For the bees, it is a critical element in hive maintenance. They use it for sealing holes and cracks, for gluing things together, for weatherproofing and as a sticky trap to dissuade ants and other creatures from entering their domain. They make it from the resin of trees. Today it is often incorporated into cosmetic and health supplements.

Pollen

Bees gather this and carry it the little sacks on their back legs. Once back in the hive, it is unloaded and used as a protein source for the growing brood. Again it is valued as a health supplement and is highly in demand.

Royal Jelly

This has become very popular in the top range of cosmetics over the last ten years. It is also used as a health supplement. The bees will produce it in an area called the hypopherangeal gland and store in empty queen cells. Although you will only be able to access it in minute quantities, it is highly sorted after, and the beekeeper with a good marketing sense can benefit greatly if they go down this route.

Bee Venom

The bee's venom can be harvested without harming the bees though this is s specialist area. The venom, which is a clear colorless liquid, is prized in alternative medicine.

You have now seen that there are many ways in which you can benefit from having beehives. On top of this, they can be used to pollinate crops, and some breeders specialize in producing nucs or queens.

Mead

At a totally different end of the spectrum to bee venom, which is a relatively new product to be marketing, you have mead. Mead is an alcoholic beverage made from fermenting honey with water. It can also incorporate other products such as herbs to add to the flavor. It ranges in strength from as little as three percent right up to twenty percent. It has been drunk since ancient times and is still popular in certain circles today. This ancient drink is surrounded by mythology and offers one more bi-product to the savvy beekeeper.

Though all these alternatives exist, honey will always be the main product from which the beekeeper can benefit. Unless you plan on heading into large-scale pollination, honey is where you will see the greatest returns and where most beekeepers derive the most pleasure.

CHAPTER 8

COMMON MISTAKES

Beekeeping is a large subject with a myriad of different aspects to it, so you are going to make some mistakes. Don't let this knock you off course or overwhelm you on your journey. The important thing to bear in mind is that there is plenty of help out there, and you only have to reach out to access it. Below are some of the more common mistakes so that hopefully you can avoid those at least. Some of them we have already touched upon, but this chapter will offer you a quick place to refer to if you want to check you are not going wrong.

Not Wearing a Suit

Many new beekeepers learn from old hands in the game. Old hands tend to have a very relaxed approach to what they wear in terms of suits, veils, and gloves. It is super tempting to try to emulate them. After all, suits are hot; they are cumbersome, and they make you look like a bit of a dork. Worse still, they advertise immediately to more experienced keepers that you are the new kid on the block. This desire to be seen to fit in probably stems back to when we the only kids with training wheels on our bike. That may have left psychological scars, but it saved you from real flesh and blood scars. Preventing stings in your formative beekeeping days means that you can focus on all the other things to learn, and as you have probably gathered from this book, there are plenty of those. Experienced beekeepers are calm, and this rubs off on their bees. You might have nerves of steel, but you will still make the clumsy maneuvers that people make when they are new to anything. This will agitate your bees, and they will sting. It is that simple. Just avoid the problem and put on the suit.

Not using enough Smoke

You use that smoker for a purpose. Firstly, just getting it alight and then keeping it alight is a bit of a knack. Practice a little before finding yourself in front of an open hive with no ammunition. Once you know how to use that smoke, then make sure you use it. If you don't, the bees are going to get angry and sting, and as we all know, that is the end for the bee. Why kill a load of your stock because you don't want to stress your stock by smoking them a little. There are beekeepers that have stopped using smokers and instead spray the hive with a spritz bottle containing water and their choice of natural herb essence. There is probably no problem with this, but you would need to be walked through it with someone who knows what to mix with the water. Have an active smoker nearby just in case.

Wrong Hive Position

Placing a hive needs to be considered carefully. There should be enough space around that you can move comfortably, and it should be on flat ground. People sometimes perch the hives on a slope, forgetting that when working it conditions might not be the same. Positioning an empty hive is easy enough, but working a full hive in damp conditions with tools and a smoker while in a suit makes things much harder. Being on a slope or in a tight space is just something you don't want to do.

Inspection Issues

This is common, and we have already looked at the importance of doing regular inspections. If you have just installed your stock and are new to the game, the temptation to keep checking on them can be quite high. Every ten days in the warmer months will keep you informed of any problems. There is nothing to stop you from doing external inspections to see how much activity there is, but too much opening of the hive is stressful for the bees. Equally dangerous is failing to check on them often enough. All of the problems that can occur can usually be rectified, controlled, or illuminated if you catch them early enough. One of the problems that tend to occur here is that the new owners go away on holiday. Peak bee inspection time does coincide with what many people regard as a holiday time. As always, the answer is to be part of an active beekeeper community of some kind. Most keepers will nor mind popping in to keep an eye on your hive once in a fortnight, and you will probably be able to repay that favor when they go away. A fresh pair of eyes might also spot a problem that you had overlooked.

Not Feeding Enough

There is nothing sadder than finding a hive full of dead bees huddles together because a beekeeper failed to supply the bees with enough supplies to get them through the winter. The reason bees don't have supplies is that we stole the honey they made through the summer. It is beholden on the beekeeper to ensure that he replaces honey with sufficient alternative food to keep that hive in a health state. Not only can the bees not fly during the cold months, but there are generally no flowers from which they can harvest either.

Taking too much Honey

This problem is linked to the problem above. In the first year, it is accepted practice to let the bees have all the honey they produce. They are getting established, and they will need a winter reserve. How much you take over the course of the second year depends on a number of different factors. Colder northern climates mean they will need more reserves, and the summers are probably shorter. Warmer climes mean that flowers will appear sooner, and the bees have a longer harvesting season. How much food they require during the lean months is something that you will become better at judging as your experience grows but always err on the side of the bees rather than your honey larder. Also, with changing climatic conditions making weather forecasting more difficult, the beekeeper can't be sure that the seasons will follow familiar patterns. Plenty of honey in the hive can only lead to healthier colonies, so think of this as long term investment over short-term gain.

Space for Rogue Comb

Bees will hang comb wherever they find space. Hives are designed so that where there is space, there is something upon which the bees can hang the comb. If you decide to leave a few frames out, that will not deter the bees from making comb, but they will do so by hanging it from the walls or the roof or anywhere else that seems suitable. When you open the hive, this breaks the rogue comb creating not only a mess but also angering the bees.

The main reason that newbie keepers tend to leave out frames is that they think either the bees don't need that many, or it will give them more room to move around when they come to inspect. This is not going to work. If you are finding it too difficult to work with the frames so tightly packed together, you can leave one out, but only one and only if you space all the remaining frames evenly. By way of creating a habit, it is probably better just to put in all the frames that the hive was designed to hold.

Starting with just one Hive

This is an easy one to understand. With so much to learn and a maintenance schedule to adopt, the logical thing is to keep the first step as small as possible. One hive might seem easier to manage, but it leaves the beekeeper with no fallback position if something goes wrong. If you lose a queen or your hive becomes too weak, there is no possibility of combining it, and you are condemning it to death. With a second hive, you have options. No two hives behave in exactly the same way, even under almost identical conditions. With two hives, you are able to compare one with another, and you will learn more with very little extra effort on your part. By you have suited up and lit your smoker, you have done much of the work required on inspection days. It makes sense just to go on and open a second hive while you are kitted up.

Not Recognizing you have Lost the Queen

This is an easy mistake for the newbie to make. Fortunately, it is quite rare to lose a queen, but when you do so, you need to take action. It is very important to point out that you do not have to actually see the queen in order to know that she is there. Providing you can see fresh, healthy eggs, then she is around somewhere. It is only when there are no eggs that you have a problem. Believing that you just must see the queen establish her presence means you may well be keeping the hive open for too long. Just holding the frame up to the light and checking that there are small white eggs at the bottom of the cells will advertise her presence. Eggs turn to larvae within three days, so you will have at least established that she was around within that period, and when you next inspect, you can check again. Even more, experienced beekeepers won't always see the queen. As your beekeeping improves and you learn to read what the worker bees are doing, that will normally show you where she is, but actually sighting her is not critical.

Not Keeping Records

We want to deal with bees, right? We don't want to get bogged down with paperwork and writing things down. Record keeping is not just going to tell you what was happening in a hive a month ago or a year ago. It is part of the ongoing learning process and will reiterate things that you may have forgotten. It gives you a season by season guide so that you know that if the bees started bringing in ivy honey in late October last year, they could well have access to that same food source this October. It lets you know the verroa count for the last year, and with that information, you will be able to assess whether it has increased dramatically this year. These and a hundred other little bits of information all add together to help you progress as a beekeeper. Keep your notebook near your bee suit and treat it as something just as important.

As a beekeeper, you are never going to stop learning. The subject is just too vast, and it is not stagnant. Techniques evolve, equipment evolves, even the pests that you deal with evolve. More importantly, the hives that you maintain and which to a very large extent are dependent on you for their very survival evolve. The process of acquiring knowledge is steep in the beginning, but soon it will plateau off into a gentler slope that becomes one of the greatest pleasures of keeping bees.

CONCLUSION

Congratulations! You have got to the end of this book and now have enough knowledge to take you on a journey that could last you for many years or even a lifetime. I imagine that there are questions that you still have, techniques you still wonder about, procedures you are not sure of. The bad news is that even if you keep bees for the rest of your life, you are always going to have questions and more things to learn. The good news is that even if you keep bees for the rest of your life, you are always going to have questions and more things to learn.

Beekeeping is not a destination. It is a constant path towards knowing more, that you will never get to the end of. In part, that is what keeps a good beekeeper going. That constant sense of adventure each time you open a hive should never go away. It is a subject that should tweak your curiosity for as long as you deal with bees. You have, after all, involved yourself in the management of nature in a way that is almost miraculous. Enjoy the trip.

BEEKEEPING

www.ingramcontent.com/pod-product-compliance
Lightning Source LLC
Chambersburg PA
CBHW052059110526
44591CB00013B/2276